ENVIRONMENTAL MANAGEMENT IN A LOW CARBON ECONOMY

Each government recognizes that there is a potential loss of competitive advantage of its business sector if future economic growth strategies are not aligned with a low carbon future. Some multinational organizations recognize this imperative and the importance of aligning business activities to a more sophisticated and flexible environmental management system that also incorporates quality, safety, occupational health and corporate ethics. An organization's Environmental Management System (EMS) has traditionally been designed to address legislative and regulatory requirements. It has now become a measure of an organization's attitude towards balancing the environmental, economic, cultural and social needs of its trading communities.

By using real world case studies this text positions the EMS as a core and critical management tool and a key requirement for businesses' long term survival. It provides fundamental building blocks to implement an EMS and clearly illustrates how it can be positioned within an organization to deliver innovative products and services to compete in a low carbon economy.

Environmental Management in a Low Carbon Economy will prepare students and professionals alike with the ability and understanding to implement an EMS which in turn will aid organizations in facilitating their transition to operate in a low carbon economy.

Stephen Tinsley is Vice Principal of Corporate Development at Edinburgh College, UK.

ENVIRONMENTAL MANAGEMENT IN A LOW CARBON ECONOMY

Stephen Tinsley

Routledge
Taylor & Francis Group

LONDON AND NEW YORK

First published 2014
by Routledge
2 Park Square, Milton Park, Abingdon, Oxon OX14 4RN

and by Routledge
711 Third Avenue, New York, NY 10017

Routledge is an imprint of the Taylor & Francis Group, an informa business

British Library Cataloguing in Publication Data
A catalogue record for this book is available from the British Library

Library of Congress Cataloging in Publication Data
Environmental management in a low carbon economy / Stephen Tinsley.—
First Edition.
     pages cm
   Includes bibliographical references and index.
   1. Sustainable development.   2. Environmental management.   3. Carbon dioxide mitigation.   4. Greenhouse gas mitigation.   I. Title.
   HC79.E5T56 2014
   338.9'27—dc23

                                                            2013030319

ISBN: 978-0-415-85548-8 (hbk)
ISBN: 978-0-415-85550-1 (pbk)
ISBN: 978-0-203-73611-1 (ebk)

Typeset in Bembo
by RefineCatch Limited, Bungay, Suffolk

Printed and bound in Great Britain by
TJ International Ltd, Padstow, Cornwall

Dedicated to
Beatrice Mary Tinsley (nee Egan)
1933–2012

CONTENTS

FIGURES

TABLES

BOXES

PREFACE

Purpose, structure and content

Many of the world's governments are striving for a low carbon economy and each views organizations, large and small, as the delivery mechanism for this ambition. For organizations to deliver on this ambition, they will need an all-encompassing environmental management system (EMS) that delivers low carbon products and services. Such a system requires the integration of quality, health and safety, social, ethical and environmental objectives into its business operations. It should also link with many stakeholders clustered around a low carbon capability, or adaptation theme at a national, sub-national and transnational level.

An EMS has, traditionally, been designed to manage an organization's environmental legislative and regulatory requirements and minimize corporate risk. These systems must now evolve to become transparent and measureable statements of an organization's commitment to produce low carbon products and services.

Today, the majority of the world's developed and developing economies are reliant on high-carbon activity. Since the publication of the Stern Review (2006), this is no longer an option for many countries. Future economic growth is dependent on low carbon policies and public and private sector organizations working together to assist new and existing organizations to adapt and develop low carbon products and services within economic constraints. The economic drivers of productivity, competition, investment, skills development, enterprise, entrepreneurship and innovation are still relevant. However, the environmental management infrastructure required to facilitate an organization's transition to operate in a low carbon economy needs to become more sophisticated.

Governments recognize that continued depletion of natural resources and the increased cost of carbon based products and services is unsustainable and that this is presenting an emerging risk to many economies. Many governments anticipate a potential loss of competitive advantage to its business sector if future economic growth strategy is not aligned with a future low carbon strategy. Some multi-national organizations are beginning to recognize this imperative and the importance of aligning business activities to a more sophisticated and flexible EMS that also incorporates quality, safety, occupational health and corporate ethics.

Key organizational partnerships are also being established to enhance environmental and ethical credentials and to mitigate risk of major international environmental crises. These partnerships are beginning to emerge as international environmental management networks with a set operating code to facilitate the trade of low carbon products and services in other countries. It is anticipated that in a short period of time these networks will have the potential to form new trading areas or clusters to exchange low carbon environmental goods and services.

The text and cases in this book explain what an EMS is and how it can be positioned within an organization to deliver innovative products and services to compete in a low carbon economy. The first part of the book briefly describes the origins of the EMS and the organizational drivers for its utilization and discusses where an organization might position itself between environmental management compliance and sustainable development at opposing ends of the environmental management spectrum. The second part of the book explores the fundamental building blocks of an EMS and provides details for implementation into an organization. Also identified and discussed are the potential organizational barriers that can arise during the implementation process. The third part of the book investigates types of EMS models, their strengths and weaknesses and how each can be customized to link with organizational structures and business strategies demanded by economic conditions and the marketplace. The final part of the book discusses whether existing environmental standards and systems are fit for purpose or whether they need to change to enable organizations to operate in a low carbon economy. This discussion considers the required changes and the potential barriers and impacts on a wider international stakeholder and customer perspective. Chapter 6, particularly, puts forward some of the environmental management and low carbon technology development issues that face organizations and countries in Asia, Europe and America. The chapter concludes with the development of an EMS that transcends international and national environmental standards. It is a system that enables engagement with institutions, agencies and organizations in sub-national, national and transnational networks, working with an enhanced code of environmental management, economic sustainability and corporate responsibility that promotes innovation and provides goods and services for the low carbon environmental goods and services sector (LCEGS).

The detailing of a new route to low carbon development helps to advance the notion that full engagement of sub-national, national and transnational stakeholders is important to move the climate change and low carbon development agendas forward. It also suggests that taking the necessary action to tackle climate change chimes with strong public consensus and is more effective in addressing local development issues, such as the provision of basic services, greater energy and food security, and employment. This book addresses options which, when tailored to specific circumstances, could help balance the pursuit of economic growth and the investment needed to transition to a low carbon economy. It also considers how to meet the adaptation needs that many organizations and countries face.

This book seeks to contribute to the ongoing discourse concerning the links between economic development and climate change, and help provide sub-national authorities with new insights and guidance as they seek to take steps to adapt to climate change. The emergence of a low carbon economy creates a dynamic environment for all organizations and their stakeholders. The constant refreshing and realigning of the Eco-Management and Audit Scheme (EMAS) and ISO 14001 standards and the increasing number of new energy efficiency and sustainability standards are testimony to this. Organizations producing products and services within a network of low carbon themes require greater flexibility to adapt quickly to change and meet the technological and innovative advances that will rapidly emerge from the knowledge exchange and collaborative investment of the networks. Such positive changes can be quickly assimilated into an organization and its network, whereas the constant changing of a standard to facilitate change only improves the standard. Additionally, the constant increase in the number of standards created to manage an organization's low carbon aspects serves to create additional sets of subsystems increasing the complexity of achieving diverse sets of objectives and targets, whereas one set of objectives positioned around a low carbon theme guides the whole organization and its supply chain to deliver products and services to an agreed set of low carbon performance indicators. The introduction of a number of low carbon themes does increase the number of objectives but it does not add new sets of guidelines or criteria and fits easily into an organization's existing set of performance measures or balanced scorecard indicators. Therefore, structure, systems, procedures and processes are structured at three levels within the organization to work with a network of organizations, agencies and institutions at the regional, national and transnational level to deliver the low carbon and economic objectives of the organization and the network.

ACKNOWLEDGEMENTS

For me, book writing is a big personal challenge and one that I know I could not complete on my own. There is no doubt in my mind that without the support of my fantastic family – Catriona, Emily, Hannah, Janey and Olly – you would not be reading this book today. They have coaxed and cajoled me over a long period and they must be greatly relieved that it is all over . . . until the next time. I would also like to thank my Dad (John) for his encouragement during a very difficult time.

Finally, I would like to give a big thank you to everyone at Taylor & Francis for all their support and guidance throughout the process.

INTRODUCTION

Summary

This chapter centres on understanding and profiling a low carbon economy, outlining its associated economic, environmental and social benefits and how to stimulate a transition from a high carbon economy. The discussion continues with the identification of the key drivers for a low carbon economy and the potential barriers. It concludes with an international perspective on how to lay the foundations for a low carbon economy and to realize the anticipated, economic, social and environmental benefits.

Context

In an austere economic climate, a world of dwindling natural resources and insecure energy sources, organizations have many demanding stakeholders operating in a range of local, regional, national and international markets. Each stakeholder expects the highest quality of product or service coupled with a growing demand for greater transparency of resource usage, energy efficiency and levels of environmental impact and social responsibility of business activities.

The demand for environmentally and socially responsible organizations is no longer a one state or one country pursuit. Many nations have committed to, or are signalling a commitment to, a low carbon economic strategy based on the development of new industry sectors. The aim is to produce low environmental impact and socially responsible products and services to create new market opportunities for businesses and a stable and vibrant economy. Increasingly, consumers of all countries are supportive of a future low carbon society and have greater knowledge of the impact of high carbon economies due, in part, to high profile issues such as increasing energy prices, global warming and climate change. Organizations, in the light of this increased consumer knowledge and multi-government agreements to deliver a global low carbon economy, recognize that merely complying with environmental legislation and regulation is not enough. Organizations now have an opportunity to embrace innovative and progressive environmental management systems that incorporate environmental and social, as well as economic, ambitions, which are clear, transparent and accessible to their customers and stakeholders. The message from governments around the world is clear, each nation must move from being a high carbon economy to a low carbon economy. Such a transition would deliver three key outcomes:

- Accelerate job creation, innovation and economic growth.
- Promote technology driven economic competitiveness.
- Increase energy security while reducing environmental impact.

The burden of creating a low carbon economy is being placed upon the shoulders of organizations worldwide.

It began slowly at first with a multi-nation commitment to the Kyoto Protocol (2005). This was quickly followed by the release of the Stern Review (2006) and the more recent commitment from the European Council (2007) to an energy and resource-efficient European economy. With the release, by many countries, of low carbon economy strategy documents, including the USA (EPA, 2007), Asia (ADBI, 2012) and UK Government (BERR, 2009), the political appetite to move from high carbon economies has rapidly and dramatically gathered momentum in a short period of time. The message is consistent, leaders of all developed nations recognize that policies and strategies should take shape through a cohesive set of international, national, and subnational low-carbon economic growth strategies; and that such strategies should focus on developing, producing, and commercializing low-carbon technologies for a healthier global economy.

There is currently no simple agreement or definition of what could be termed a low carbon economy. However, following publication of the Stern Review (2006) many countries pledged a commitment to a low carbon future and delivered strategy documents envisioning a low carbon economy based on their commitment to developing products and services based on low carbon technology development.

The increased frequency of climate change and other environmental degradation events has not had the desired effect in terms of accelerating the need to properly finance low carbon technology development to mitigate and adapt to these events. However, combined with the prospect of economic growth particularly through a low carbon environmental goods and services economy, low carbon technological development is becoming increasingly attractive and urgent for many countries. Current, historically informed wisdom states that global-scale environmental challenges can only be met through technological innovation supported by private and public sector investors and policymakers. In pursuit of this future, the drive towards technological development goes beyond the traditional private and public categorizations. Assessing the effectiveness of European, North American and Asian private sector funding instruments and strategic initiatives may identify ways that public investment or other policy actions can leverage financial instruments to address market failure or urgent societal need and draw on comparisons of international financing best practice.

Theory of a low carbon economy

An economy consists of a country or recognized area, where the resources of labour, land and capital are utilized by economic agents to facilitate the production, exchange and distribution of goods and services for consumption by its citizens. The majority of developed and developing country economies are considered to be high carbon economies. This is as a result of increased utilization of fossil fuel energy sources such as oil and gas by the agents for economic growth, releasing high levels of greenhouse gas (GHG) emissions into the environment. A low carbon economy, in contrast, is an economy based on low utilization of fossil fuels and consequently releases a minimal amount of GHG emissions into the environment; carbon dioxide gas (CO_2) is by far the largest component of the harmful GHG mix. A high carbon economy has little or no restraint on the carbon intensity of energy production and energy consumption or environmental pollution. A low carbon economy targets low energy consumption, low pollution, low emission and high technology, high energy efficiency and strong economic benefit (World Low Carbon and Eco-economy Conference and Technical Exposition, 2009).

Recent opinion, held by scientists, industry practitioners and the increasingly knowledgeable general public, centres on the belief that there is an increasing accumulation of harmful GHGs, particularly CO_2, in the atmosphere due to the high carbon economic activity and the lifestyle choices of its citizens. Arguably, the consequences of such high carbon activity are presenting themselves in the form of more frequent and unusual weather events, reinforcing the opinion that the climate is changing. Further credence is therefore given to the assumption that as increased concentrations of harmful greenhouse gases accumulate in the atmosphere so weather events will become more extreme thus producing long-term climate change activity, with negative and, it should be said, positive impacts on economies and citizens in the foreseeable future.

The implementation of a global low carbon economy, therefore, is being proposed and accepted by many governments as a means to reduce occurrence and impact of future catastrophic climate events and to act as a road map to the realization of a low carbon economy based on sustainable and secure energy sources.

All developed countries, together with many developing countries, are considered to be high carbon economies. In response to the growing bank of evidence, these developed and developing countries have signalled their intent to move towards the implementation of a low carbon economy. However, the financial sums involved in making this economic transition are considered to be daunting and many informed publications from respected organizations point to this fact. At present concentrations of GHGs in the atmosphere equal 350 particles per million (ppm). The International Environment Agency (IEA) estimates that limiting GHG concentrations to 450 ppm CO_2 equivalent would require US$550 billion to be invested in low carbon energy technology development from now until 2030. The United Nations Development Programme (UNDP) estimates the cost of climate change adaptation at US$86 billion. The Stern Review (2006) recommends that governments should act now and invest the equivalent of 1 per cent global gross domestic product (GDP) to address climate change and warns that the 'failure to act' option would cost the equivalent of 5 per cent of global GDP. These publications also state that, in the coming years, the majority of financing will need to come from private sources, or from innovative funding mechanisms currently available or under development. The existing national financial commitment of Overseas Development Aid (ODA), while significant in stark monetary terms, is unlikely to be sufficient to finance the necessary technological investments. To put this into context, for energy-related development activities the ODA, at present, provides US$5–7 billion per year, based on the IEA figure of US$550 billion required to be invested in low carbon technology development. The Overseas Development Fund represents just 1 per cent of the total amount required (UNDP, 2009).

Direct technological investment to aid climate change adaptation and mitigation on this scale, in the current weak global financial period is, some would say, not an option for many developed and developing nations. Indirectly, however, the international community is currently trying to provide assistance by piloting a number of public policies, strategies and new market-based, innovative financial instruments and mechanisms, to attract and drive investment towards low carbon and climate change adaptation technologies and practices. In 2007, the private sector invested nearly US$150 billion of new money into low carbon and clean energy technologies in response to these new policy and financial incentives. There exists some concern that the current global financial crisis may freeze financing for low carbon technology projects and that a number of financial incentives to support energy efficiency and renewable energy technology development will be phased out by governments striving to trim budget deficits. However, it is expected that investment in low carbon technologies will resume its growth to about US$450 billion in 2015 and US$600 billion by 2020 (UNDP, 2009).

The financial investment options on offer, however, often remain inaccessible for many OECD (Organization for Economic Cooperation and Development) countries as well as a small number of rapidly developing countries. Such barriers to funding need to be removed before technological investment schemes can be widely disseminated and easily accessed by developing countries. For example, the Kyoto Protocol (2005) created the Clean Development Mechanism (CDM) to promote both sustainable development and GHG emission reduction in developing countries. The CDM is a global cap-and-trade mechanism, which enables developing countries to earn credits for their emission reduction projects and to sell these cheaper credits to industrialized countries. Despite its low carbon technology development potential, there is strong concern that only a limited number of countries will be able to benefit from the CDM, and that this, and potentially other, financial investment mechanisms could bypass countries in Africa entirely because they are starting from such a low technological base.

At present, five countries – China, India, Brazil, South Korea, and Mexico – are expected to generate over 80 per cent of all CDM credits by 2015 leaving just 20 per cent for all other countries. These rapidly developing countries could be considered to be in advanced transition from developing to developed country status and are therefore more attractive for developed countries, financial markets and investment funding bodies to work with. Existing financial market regulation all too often fails to attract investors into low carbon technologies and sustainable energy projects. The specific market conditions of developing countries need to be incorporated into the design of new innovative financial mechanisms. A number of reforms to the CDM are currently being discussed to achieve this objective. Simultaneously, developing countries need assistance to put into place an enabling environment (e.g. public policies, institutions, human resources) to be better placed to access these new sources of finance.

In addition to the difficulties in accessing finance, the sensitivity around the global pursuit of low carbon technologies to solve national economic, social and environmental issues is that 'one size does not fit all'. Developing countries, with greater social and environmental issues, need to place a larger part of their potential low carbon technology development budget towards assisting social and environmental development programmes, thus a smaller investment fund is available for technological development. The choice then for these countries is to either fall further behind in technological development and global economic development terms than their developed country neighbours or neglect the social and environmental welfare of their citizens while delivering economic growth.

A low carbon economy can mean different things to different countries depending on the economic, social or environmental objectives to be achieved, thereby making it very difficult to clearly define. The majority of global citizens will have limited or no experience of a low carbon economy. Citizens in countries such as Sweden, Norway, Finland and Canada may point to one or more of their Sustainable Municipalities as small, representative models of what a low carbon economy may look like. Sweden, led by its Commission on Oil Dependence (2006), aspires to be free of oil as an energy source by 2020.

Theorists tell us what a low carbon economy would look like if certain renewable energy and resource efficiency changes are introduced and a low carbon lifestyle is adopted by its citizens. In defining a low carbon economy, it is easier to say that it is an economy that emerges at some point in the future where the planned transition from its current high carbon economy is recognized by society as having achieved a balance of economic, social and environmental activity that has an acceptable impact upon the environment. It is more difficult to say which path each country should take to achieve such an economy. Historically, a view of 'future' economies has been determined, largely, by technological development and the balance of economic, social and environmental outcomes has been heavily weighted in favour of economic growth. Realistically, it is unlikely that this pattern will change, but it is hoped that the balance between economic, environmental and social outcomes will be more evenly weighted.

Concern about climate change and dwindling natural resources are key drivers for those nations wanting to work towards a low carbon global economy. Many countries are taking action to mitigate and adapt to the challenges thrown down by a changing climate, and the forewarning expressed in the Stern Review, to move towards a low carbon economy. These countries are, again, beginning to invest in technologies which, it is hoped, will deliver a low carbon industrial global economy. With this investment, each country will endeavour to ensure its businesses are well placed to engage with emerging opportunities and assist the transition to an economy that is based on efficient resource use and fuelled by secure, renewable energy sources.

A low carbon economy is, therefore, viewed by many governments as an economic model focused on the delivery of new business opportunities and on the minimal consumption of high-carbon energy sources such as coal, oil and gas and the minimization of the resulting output of harmful GHGs, specifically CO_2. An economy based on high-carbon activity is no longer an option for many countries. Future economic growth will be dependent upon low carbon policies and public and private sector organizations working together to assist new and existing businesses to adapt and develop low carbon technologies within economic constraints. The economic drivers of productivity, competition, investment, skills development, enterprise, entrepreneurship and innovation are still relevant but the existing technology development and investment infrastructure needs to change to facilitate the transition to a low carbon economy as follows:

- Competition – Strong and open competition provides the basis for efficiently working markets where businesses and investors have the incentives to take advantage of the opportunities offered by the transition to a low carbon economy. They also provide strong incentives for firms to improve their efficiency, deliver better quality goods and services, and invest in innovation.
- Investment and skills – Greater investment in physical capital and human capital to allow firms to improve their internal efficiency and exploit new technologies and opportunities from a low carbon economy.
- Enterprise and entrepreneurship – New business start-ups and growth are important in new and growing sectors, such as the low carbon and environmental goods and services (LCEGS) sector. Entrepreneurs enable innovative ideas and technologies to enter the market swiftly. They provide greater competitive pressures to those within the market to adapt and become more efficient.
- Innovation – The progress of some sectors or subsectors in making the transition to a low carbon model will depend on the extent of innovation and evolution of technologies and production processes. This also refers to wider innovation across the economy in terms of adopting more energy efficient processes and

day-to-day behaviours. The Carbon Trust (2008) has highlighted the importance of innovation to establish low carbon activity in the UK. In addition, it identifies a further three drivers of industrial transformation:

○ Consumer behaviour – Changes in consumer behaviour and preference affect levels and growth of demand for new (low carbon) products and can influence the move to more resource-efficient industrial processes.

○ Cost of carbon – The establishment of a robust price for carbon and the continued growth of carbon trading markets can internalize the true environmental costs of emissions, expose competitive differences between operations and should drive innovation in technology and measures to reduce pollution over the longer term at the lowest cost.

○ Targeted regulation – The regulatory framework can have a significant impact on business decision-making. Rigorous assessment of the need for, and impact of, policy options is required to ensure that measures introduced do not have a disproportionate impact on business costs and do not stifle investment incentives or the adoption of new technologies. Many governments have set out long-term regulatory and policy frameworks to provide clear and credible long-term signals to businesses to help shape investment decisions, principally through participation in the European Union Emissions Trading System (2009) to establish a robust carbon price, through the Renewable Obligation (RO), to incentivize investments in renewable energy and also through the Climate Change Act (2008) which commits the UK to statutory targets for emissions reductions. The policy framework also seeks to remove barriers (including those resulting from information failures) to investment in new technologies, innovation and the human capital required to create a low carbon economy.

Businesses also require clarity on the availability of, and access to, business support policies, including access to information on low carbon opportunities and measures, how to calculate their carbon impact and where to get appropriate financial help from central and regional funding sources. In order to make grants, subsidies and advice more targeted and easier to access, governments have introduced a range of business support simplification initiatives. For example, in the UK this has resulted in the reduction of hundreds of publicly-funded support schemes to a single transparent, accessible portfolio of products under the support solutions for business banner.

Through new energy development, technological innovation and industrial restructuring, a low carbon economy will produce a better balance of economic, social and environmental drivers. In short, a low carbon economy can be characterized according to four aspects (United Nations, 2009).

• All waste to be minimized and a greater emphasis placed on waste reduction, repair, reuse and recycling.

- Energy production and consumption should utilize low carbon energy sources and energy efficiency methods together with renewable and alternative energy sources, carbon capture and sequestration.
- Higher efficiency of energy utilization. All energy resources should be used efficiently, with more efficient energy conversion devices and combined heat and power applications.
- There should be high awareness and compliance at all levels of the economic community with environmental and social responsibility needs and initiatives.

Key drivers for a low carbon economy

A significant array of public policy guidance and financial instruments is available to help many regional authorities implement new technology and business development measures in industry sectors. However, on their own, policy guidelines and financial incentives are rarely enough. This is best explained using wind energy as an example. Although wind technology is a rapidly growing and maturing technology, it is, on its own, not sufficient to ensure its widespread development and acceptance as a significant alternative to energy cost reduction. Only countries that have established an enabling environment, comprising stable comprehensive public policies, tools to manage externalities (e.g., noise and landscape for wind energy) and public consensus and acceptance, strong political commitment and adequate access to financing have succeeded in tapping into its power as a major energy source. Such commitment in countries and regions focuses not only on reducing costs and improving revenues to increase profitability, but also on reducing risks. Once priority technological and non-technological options have been selected the next step is to convert these options into a mix of optimal policy instruments that fit with a set of legacy-focused, coherent projects. Decision-makers at the local, regional and national levels need to:

- Understand the specific workings of key sectors and markets in order to design an appropriate mix of policy measures for priority low carbon technology development activities, including combining and customizing relevant financial instruments.
- Translate policy measures into a synergistic set of initiatives to identify and sequence projects that lead to organizational strengthening, and support of individual technological projects.
- Match technology development opportunities with the most appropriate and available sources of funding such as ODA, World Bank and market-based mechanisms such as CDM and feed in tariffs.
- Develop the required documentation and prepare to meet the due diligence requirements that are unique to each source of funding and new technology development needs.
- When the chosen technology and relevant funding option leads to increased pressure on one or more of the partners, such as public budgets, individual

developers must be prepared to reduce these pressures by developing innovative and flexible financing instruments.

The Montreal Protocol (1998) advocates for the acknowledgement of the policy and implementation competencies of regional governments, related to a range of sectors which both directly affect and are impacted by climate change. The Signatories to the Protocol also committed to actively participate and take action in future international climate change endeavours, in line with the principle of common but differentiated responsibilities and respective capabilities. A series of arrangements for twinning regions will provide a conduit for exchanging information and best practice experiences. Much can be learned from these experiences on how to scale up efforts to progress global low carbon technology development.

Case study

The state of California (USA) hosted the Governors' Global Climate Summit (2008), which brought together regions from China, India, the United States, Canada, Mexico, Brazil and Indonesia. Participants committed to work together by signing a joint declaration to forge partnerships in the areas of forestry, cement, iron, aluminium, energy and transportation and to focus research, development and deployment activities on areas such as energy efficiency, renewable energy and zero- and low-carbon electricity generation and fuels. Exchanges are already well underway.

Building on theses international activities and the Montreal Declaration, a 'statement of action' was delivered to the United Nations Framework Convention on Climate Change (UNFCCC) (2008) in Poznan, Poland, on behalf of regions from around the world and supported by The Climate Group. The 'statement' contained new commitments by subnational governments, from setting energy efficiency and renewable energy targets to specific exchanges in the areas of efficiency, renewable energy, clean transport and land-use, partnering with developing country regions through the UNDP, and supporting the work of the World Summit of Regions and the Governors' Global Climate Summit (2010).

Laying the foundations for a low carbon economy

The commitment to achieving a low carbon economy is the same for the US, Asia and Europe but the paths to its realization may be different for each region.

Establishing a low carbon society would, arguably, appear to be more urgent in Asia where GHG emissions are increasing rapidly due to high economic growth and increasing demand on energy and natural resources. Although traditionally Asian societies have adopted many low carbon pathways of development including frugal lifestyles and population control, current trends and projections identify future development patterns that are still associated with a large carbon footprint.

It is practically impossible, due largely to land restrictions, for many developing Asian countries to follow the same historic growth patterns as US and European countries; there is therefore a need to consider alternative growth models for a low carbon society that also meets the economic and cultural needs of Asian countries. In a resource efficiency assessment report issued by the Institute of Global Environmental Studies (2011) several stakeholders stressed that the design of the future economy should aim to change energy-intensive lifestyles and consumption patterns, and consider a new set of carbon standards to promote such a transition in all countries.

The core of an Asian low carbon economy may well centre on energy efficient technology development, clean energy structures and the development of sophisticated information and communication technology (ICT), grid technology systems. A 2010 report by the National Institute of Environmental Studies (NIES) on Japan suggests that the 20 per cent reduction of global GHG emissions by 2030, and a 50 per cent reduction by 2050 or even 80 per cent by 2100, is possible provided rapid transformation of social, industrial and economic systems takes place in the medium to long term. For example, a 70 per cent reduction of CO_2 emissions by 2050 (compared to 1990) is feasible in Japan if a 40 per cent reduction in energy demand is combined with a decarbonization of the energy supply. Reductions in energy demand of 20–40 per cent in industry (through structural changes and introduction of energy efficiency technologies), 80 per cent in passenger transport (through appropriate land use and energy efficiency improvement), 60 per cent in freight transport (through controls on the distribution system and improved energy efficiency of cars), 50 per cent in the residential sector (through high thermal insulation housing) and 40 per cent in the commercial sector are plausible.

The expected cost of introducing these enabling technologies is calculated to be approximately 1 per cent of Japan's GDP in 2050 (NIES 2010). The same study found that the introduction of an emissions trading system (ETS) and a carbon tax would not be enough to achieve a low carbon economy in Japan. It is clear that the identification of the appropriate model, together with the alignment of policies and measures, is essential if an effective low carbon economy is to be achieved in Asia. Other models for a low carbon economy may cluster around key natural sustainable energy sources (e.g. Norway, Iceland) where hydropower or geothermal power is a major source of electricity. Iceland, for instance, intends to become the world's first hydrogen economy by 2050 (Icelandic Energy 2006).

To visualize similar low carbon economy models in the Asian context, national energy strategies need to be based on a thorough reassessment of alternative energy potential through a comprehensive inventory of natural resource endowments. Most Asian countries, however, have yet to map the full potential for wind, solar or geothermal energy sources and have made only limited efforts to exploit such energy sources. In this light, the recently announced 'Cool Earth Promotion (2008)' initiative by the Government of Japan, which calls for the development and dissemination of 21 specific innovative technologies by around 2030, and a global goal of improving energy efficiency by 30 per cent by 2020, can contribute greatly to the achievement of a low carbon economy in Asia.

In Europe, energy efficiency gains in transport, industry and building sectors, decarbonization of power generation through increased deployment of renewable sources, natural gas, and coal with CO_2 capture and storage, and increased use of renewable sources of energy including biofuels for transport, are some of the measures identified to move towards a low carbon economy. Similar policies and measures could be assessed for their potential deployment in developing Asia countries depending on national circumstances.

Reducing global emissions by 50–60 per cent by 2050 at acceptable costs will require innovation in science and technology to make clean energy technologies more efficient and affordable. As deploying technologies such as solar, wind, biofuels, hydrogen and carbon capture and storage will be most crucial in Asia, technology development partnerships can be formed through the infusion of public funds. Stern (2006) recommended doubling the aggregate amount of public funds devoted to energy research and development (R&D) to reach about $20 billion per year.

Strategic regional cooperation, through effective investments, policies and measures to improve energy efficiency and promote renewable energy, will play a key role in establishing a low carbon economy in Asia. To encourage a shift in the direction of energy efficiency and renewable energy sources, greater attention should be directed to bilateral and multilateral development assistance. The role of developed countries such as Japan and other G20 economies and multilateral financial institutions such as the World Bank is crucial to accelerating the transition to a low carbon economy. Leveraging such investments with private resources is also essential. Developing Asian countries receive substantial bilateral assistance for energy, with 14 Asian countries among the top 20 recipients of bilateral development assistance for energy.

The pursuit of economic growth has been the main objective for many governments and driven, largely, by technological development. This historical model of technology driven economic development may suit European and North American countries but it may not be the ideal economic development model for Asian countries as they battle with large populations, constrained resource limitations, and now the global challenge of climate change mitigation and adaptation. So far, however, Asian governments have yet to offer an alternative economic future as many are still considering options that provide a solution to combat poverty and improve living standards for its citizens while simultaneously delivering a low carbon, energy secure and climate resilient economy. Asian countries need to become more involved in the low carbon economy debate, if only to ensure that they are aware that there are several routes to a low carbon economy and incremental steps in technology development, as opposed to giant strides, could deliver an economy based on energy efficiency and sustainable lifestyles.

Governments around the world are beginning to implement strategic frameworks to lead and facilitate the transition to a new low carbon economy. This transition is being sought to address the existing climate change issues and to deliver a new economic philosophy, a philosophy based on a low carbon industrial revolution designed to introduce new work and lifestyle patterns that have less impact upon the environment. Linking economic development with climate change remediation provides an elegant solution particularly in the current economic circumstances. It is hoped that nations will emerge from this period of severe economic turbulence on a trajectory charting increased economic opportunities and a society committed to a low carbon lifestyle.

1

THEORY OF ENVIRONMENTAL MANAGEMENT

Summary

In order to understand the concept of environmental management, this chapter provides an introduction to the development of environmental standards and the drivers that encourage corporate executives to implement environmental management systems (EMSs) into organizations to minimize reputational and financial risk of environmental incidents. A range of environmental issues and their potential impact on organizations are explored, together with the imperatives that require managers to adopt EMSs and reduce levels of corporate environmental risk. The chapter concludes with a discussion on where organizations might best position themselves using an environmental management continuum that ranges from a position of basic compliance at one end to the pursuit of sustainable development at the other.

Environmental management issues and risks

It can be argued that modern business roots of environmental management can be traced back to the 1960s and the utilization of harmful pesticides in agricultural activities. Rachel Carson, in her publication *Silent Spring* (1962), identified the impact business activity was having on the environment. She expressed her concerns at the harmful chemicals being used by the American agricultural industry in the pursuit of maximizing food crop yields and the impact of the arising pollution on the environment (Box 1.1).

Due to the increasing severity of penalties for environmental management failure (Box 1.2), large organizations were beginning to adopt environmental management systems that focused on the 'management' of environmental risk.

Box 1.1 Silent Spring

. . . pheasants, red legged partridges, wood pigeons, doves, green finches, chaffinches, blackbirds, song thrushes, skylarks, moorhens, bramblings, hedge sparrows, house sparrows, jays, yellow hammers, carrion crows, hooded crows, gold finches and sparrow hawks – **tens of thousands** of these birds were found dead or dying in the UK countryside in Spring 1961.

Box 1.2 Corporate environmental disasters

BP Gulf oil spill (2010)

The Gulf oil spill began April 22, 2010 and leaked an estimated 206 million gallons into the Gulf of Mexico, making it the worst oil spill in US history and the largest accidental oil spill in the world. Oil washed ashore in all of the Gulf States, creating health threats for both humans and animals.

The spill began when an oil well a mile below the surface of the Gulf blew out, causing an explosion on BP's Deepwater Horizon rig that killed 11 people. Oil flowed into the Gulf for 87 days before the well was finally capped on July 16, 2010.

Now ranked as the largest offshore oil spill in US history, the BP spill resulted from the April 20, 2010 Deepwater Horizon drilling rig explosion. Hundreds of millions of gallons of oil have been spilled to date and it continues to damage marine and wildlife habitats along with the Gulf's fishing and tourism industry.

Jilin City (2005)

A chemical plant exploded in Jilin City in China in November 2005, polluting the Songhua River with an estimated 100 tonnes of pollutants containing benzene and nitrobenzene entering the water. Ten thousand residents were evacuated.

Al-Mishraq fire (2003)

In Iraq, the Al-Mishraq sulphur plant, near Mosul, caught fire on June 24, 2003. The fire, which burned for about a month, released 21,000 tonnes of sulphur dioxide into the atmosphere each day. Many people were hospitalized and most of the area's vegetation was destroyed as a result. Sulphur dioxide can be harmful to people, causing serious respiratory problems, and it also creates acid rain which can destroy crops and contaminate water courses.

Baia Mare (2000)

On January 30, 2000, a dam restraining water from a gold-mining operation in Romania, in the town of Baia Mare, broke. The water was contaminated with 55–110 tonnes of cyanide and other heavy metals and travelled through several rivers in Romania, Hungary and Yugoslavia, eventually reaching the river Danube. Massive amounts of fish and aquatic plants were killed and up to 100 people were hospitalized after eating contaminated fish.

Union Carbide gas leak (1984)

On the night of December 2, 1984, the Union Carbide pesticide plant in Bhopal, India began to leak methyl isocyanate gas and other poisonous toxins into the atmosphere. Over 500,000 people were exposed and there were up to 15,000 deaths at that time. In addition, more than 20,000 people have died since the accident from gas-related diseases.

Environmental management systems and standards development

Environmental management requires policy commitments, as a minimum, to comply with all legislative and regulatory requirements and other mandates, for the prevention of pollution, and for the continual improvement of an EMS. It also requires adherence to a detailed overall structure and documentation template that can be verified by an external auditor, including an environmental policy, environmental plan, implementation procedures and responsibility assignments, monitoring, corrective action and management review procedures. For corporate environmental management to have meaning, all standards and systems should have substantial similarities across organizations, sectors and national boundaries.

The substantive details of an EMS, however, including the scope of the EMS, the pace of its improvement, prioritization of impacts, and specific choices of objectives and targets for improvement, are left to the discretion of the organization. Depending on the business decisions, the EMS in practice may be used to maintain or improve regulatory compliance, to increase efficiency in non-regulated environmental performance outcomes (such as use of energy, water and materials), to promote sustainable practices (Gallagher, 2004), or for other purposes, such as increasing employees' environmental awareness and morale, standardizing and integrating management procedures, or simply promoting a strong environmental image to business customers and the public. Internally, the EMS may also serve to spread the responsibility for environmental management across all managers whose actions affect environmental and social aspects, rather than just the environment, health and safety of staff.

What is environmental management?

The concept of environmental management emerged in the 1970s when organizations began to manage the risks associated with industrial pollution and the growing number of legislative and regulatory controls being introduced by governments. Environmental management at this early stage was seen by industry, at worst, as a fad and at best as an extension of product quality assurance. It was viewed as being a restriction on production and corrective responses from management were seen as reactive, an 'end-of-pipe' activity utilized to remedy impacts, rather than proactive and anticipatory to the potential impact of business activities.

There was early management recognition, in manufacturing organizations particularly, that pollution was caused by a variety of waste discharged from production activities. It was also an observation that reducing this waste output meant a reduction in the cost of production, a reduction in pollution and thereby a reduction in corporate financial and reputational risk. Total quality management principles were then applied together with a commitment to continuous improvement in production processes, manufacturing technologies and resource usage to minimize waste and reduce pollution. Managers began selecting strategies that incorporated design for environmental efficiency in the manufacture of products. This reduced the amount of raw materials used in the production process and optimized the reuse and recycling of any generated waste (Khanna and Kumar, 2011).

EMSs have existed in various forms since the 1970s, typically to assure regulatory compliance, reduce liability, identify waste minimization opportunities, and more generally to manage the environmental impacts of a business's activities (Andrews et al., 2004). Momentum increased again from the mid-1980s, particularly with the publication of the Brundtland report (1987) on sustainable development. Since the mid-1990s their use has accelerated worldwide, spurred by the 1996 publication of the ISO 14001 international environmental standard and similar standards in the United Kingdom and European Union. Each succeeding decade witnessed the increase in the adoption of EMSs as shown in Table 1.1.

The introduction of the ISO standards initially was to ensure compliance with rapidly increasing environmental legislation in the United States of America. At the same time, innovative organizations in Europe were developing a more proactive attitude towards environmental risks and viewing them as business opportunities. Many organizations were now fully aware of the environmental and social issues and risks associated with their business activities. They began searching for workable models to respond to the environmental and social risks and exploit the potential opportunities. Some organizations began to develop proactive strategies, determine the business case metrics and develop technology to turn environmental and social risks into business opportunities (Bekefi and Epstein, 2008).

Today, the Kyoto agreement (1987) is seen as having given birth to many government backed environment protection agencies, together with detailed national route maps to economic sustainability. Many more formal environmental

TABLE 1.1 ISO 14001 certification leading countries

	Country	ISO 14001 Certificates 2011
1	China	81,993
2	Japan	30,397
3	Italy	21,009
4	Spain	16,341
5	United Kingdom	15,231
6	Republic of Korea	10,925
7	Romania	9,557
8	France	7,771
9	Germany	6,253
10	United States of America	4,975
	Worldwide Total	267,457 Certificates

Source: ISO Survey 2011

protection agencies were established together with some informal institutions including the World Wildlife Fund (WWF) and Greenpeace.

The global economy has rapidly evolved through different stages of economic development. A long industrial past of ignoring the environmental impact of economic activities has been replaced by recognition of the need for environmental protection, corporate social responsibility (CSR) and sustainable development. These environmental and social concepts are now high on the future business development agenda of governments, organizations, and individuals in many developed and developing countries.

Standardization

In order to stimulate, monitor and measure corporate environmental and social improvement national and international standards were introduced (Brunsson and Jacobsson, 2000). In a global economy, without standardization and its results, technical standards or specifications, interchanges would become difficult at best. Standardization, therefore, sought to stimulate and regulate international trade by eliminating obstacles arising from different national practices. Standards were therefore seen as important and necessary for the promotion of economic efficiency as they provided a basis for reducing information-related transaction costs (Nadvi and Waltring, 2004).

However, as such standards were viewed as not being truly global they constituted, in many cases, non-tariff barriers for international trade. While other economic tariff barriers were increasingly being lowered so these non-tariff barriers (i.e. increased technical standards and recycling and disposable regulations affecting the requirements for products, services and, indirectly, production processes) were acquiring increasing importance (Blanco and Bustos, 2004). In short, the importance of international trade to the global economy has grown dramatically in the last two decades, but while tariffs and quantitative restrictions on trade have been lowered or eliminated, barriers of a different nature have had an increasingly restrictive effect on trade, especially in the case of a broad range of technical and environmental standards (Giovannucci and Ponte, 2005).

The British Standards Institution (BSI) defines an EMS as 'part of the overall management system that includes organizational structure, planning activities, responsibilities, practices, procedures, processes and resources for developing, implementing, achieving, reviewing and maintaining the environmental policy' (BS 8555, 2003). It can be seen as a risk management tool also – since the standards to which organizations are assessed are not prescriptive, they provide the framework within which an organization must identify its own potential impacts upon the environment, and seek to control these as well as to identify opportunities to reduce the associated risks. An EMS follows the Deming cycle of 'Plan – Do – Check – Act'.

Plan

The purpose of an EMS is to deliver the commitments stated within an environmental policy. The policy should commit an organization to legal compliance and continual improvement (BS 8555 – Phase 1, Stage 3; ISO 14001:2004 – clause 4.2). An organization should identify all applicable environmental aspects that may create an environmental impact and also all relevant environmental legislation. Compliance with legislation must also be demonstrated in order to meet the requirements of BS 8555 – Phase 2, Stage 1–5 and ISO 14001:2004 – clause 4.3.2 and 4.5.2. Additionally, in relation to environmental performance, an organization makes a commitment to continually improve, and should set out objectives and targets and a plan to achieve these (BS 8555 – Phase 3, Stage 3–7; ISO 14001:2004 – clause 4.3.3). The objectives and targets provide the organization with an improvement plan specifically tailored to ensure improvements in environmental performance related to its environmental aspects (BS 8555 – Phase 3, Stage 1 and ISO 14001: 2004 – 4.3.1) associated with the range of activities it carries out, the product(s) manufactured or the service(s) provided.

Do

Control measures and procedures are intrinsic to the successful implementation and maintenance of an EMS, whether this is related to training (BS 8555 – Phase 1, Stage 6 and Phase 4, Stage 2 and ISO 14001:2004 – 4.4.2), operational control (BS 8555 – Phase 3, Stage 6 and ISO 14001:2004 – 4.4.6), or emergency preparedness and response (BS 8555 – Phase 4, Stage 5 and ISO 14001:2004 – 4.4.7). Most organizations have a clearly defined, written set of procedures outlining what needs to be done in order to ensure compliance with its EMS.

Check

A successful implementation programme should be reviewed within the EMS through a programme of internal audits (BS 8555 – Phase 5, Stage 1 and ISO 14001: 2004 – 4.5.5) and identification of corrective and preventive actions to deal with identified non-conformances (BS 8555 – Phase 5, Stage 2 and ISO 14001:2004 – 4.5.3). Conformance of an organization against the procedures is evaluated during the audit process. Compliance with legislation is also checked during auditing as a number of operational procedures are drafted to ensure legal compliance of business activity as a minimum.

Act

The results of the 'checking' stage of the process should be used to inform management of areas of progress, or problems within the system as part of the management review (BS 8555 – Phase 5, Stage 3 and ISO 14001:2004 – 4.6). This management review process should, in turn, be used to re-address the issues of environmental policy, the organization's environmental aspects and impacts, legal compliance and objectives and targets. The continually improving organization is, in turn, more able to adjust to those changes that have occurred and reset objectives and targets for future years. According to a study conducted by Business in the Community (Environmental Index Report, 2006) an EMS helps an organization to improve its environmental performance. Common requirements of an EMS (such as a policy, objectives, targets, training and reporting) demonstrate a commitment to incorporating environmental issues into key practices with the resultant benefits of:

- Improved risk management.
- Reduced liability costs.
- Increased competitive advantage.
- More employee involvement.
- Improved public image.

History of environmental management systems

In an attempt to shift the corporate perception that environmental management is a drain on profits to one that demonstrates that environmental protection and corporate profitability should be intertwined, the Brundtland report (1987) offered a route towards sustainable development encompassing a more holistic view of the relationship between economic development, human rights, social development and environmental impact on human settlements. The United Nations Conference on Environment and Development, held in Rio de Janeiro in June 1992, became known as the Earth Summit. This summit, attended by the leaders and heads of state of 170 nations, was the blueprint for implementing the concept of sustainable development. Based on 27 principles, this blueprint became known as the Rio

Declaration. Also, emerging from this summit were environmental management standards identified to guide organizations in their efforts to implement EMSs and establish appropriate performance measures.

In March 1992, the first formal environmental management standard, BS 7750, was developed by the BSI. The standard was based on a two-year pilot programme with 230 organizations that participated by implementing and testing the standard. The responses from the pilot programme were used to modify the standard and it was finally published in January 1994. At the time of the BS 7750 standard development, the European Commission was setting out its proposal for an eco-audit scheme. Following initial proposals and amendments, the European Commission published what has become known as the Eco-Management and Audit Scheme (EMAS). The EMAS standard was adopted by the Council of Ministers on June 29, 1993, and became open to organization participation in April 1995. The development of the ISO 14001 standard came from the call for improved international environmental performance expressed at the United Nations Conference on Environment and Development (1992).

The International Organization for Standardization (ISO) was charged with creating an internationally recognized EMS (Bansal and Bogner, 2002). The ISO 14001 standard was developed in under three years. It was quicker to develop than other international standards as it relied heavily on the content of BS 7750 for a working framework (Schaltegger *et al.*, 2003). The Comité Européen de Normalisation (CEN), a European standards body, was established in 1961 to develop and manage industry standards for members of the European Economic Community (a forerunner of the EU). It was quickly realized within CEN that some member organizations would prefer to meet the EMS requirements of international standard ISO 14001, an equivalent standard to EMAS, the European standard. However, the adoption of ISO 14001 as a European standard meant that existing national environmental management standards such as BS 7750 had to be withdrawn. A consequence of this outcome was that BS 7750, so influential in the development of ISO 14001, was withdrawn in March 1997.

This resulted in two EMS standards being available to organizations in European countries; the international standard ISO 14001 and the European EMAS scheme. The EMAS regulation was revised in April 2001 and based on ISO 14001; the principal benefit of this was that it was easier for ISO 14001 certified organizations to progress to what is considered the more rigorous EMAS regulation (LRQA, 2004).

Environmental standard BS 8555 was launched after an initial pilot programme named Project Acorn. Through Project Acorn, the implementation of an EMS was trialled between October 2001 and March 2003, supported by the BSI (Sheldon, 2003). The project was funded in the main by the UK Department of Trade and Industry (DTI), and supported by the UK Department of Environment, Food and Rural Affairs (DEFRA). On completion of the trial, Project Acorn was withdrawn and the new British Standard BS 8555, the phased approach to the implementation of an EMS, was released in April 2003 (The Acorn Trust, 2003).

Organizations are becoming more proactive in managing and monitoring the environmental and social impact of their business activities. They are increasingly committed to integrating environmental and social concerns into their daily business activities. The adoption of a formalized EMS, voluntary or certified, is seen by increasing numbers of organizations as the most effective way of managing and minimizing environmental risk, minimizing resource use and allocating responsibility for targeted outcomes (González-Benito and González-Benito, 2005; Nishitani, 2011).

An EMS has been defined by the BSI (1994) as 'The organizational structure, responsibilities, practices, procedures and resources for determining and implementing an environmental policy.'

Safety audit

As a response to growing political and public pressures on organizations to minimize the environmental impact of their business operations, more organizations are implementing EMSs and conducting environmental audits to reduce the level of risk associated with environmental incidents. The first attempt at insuring against environmental incident was the safety audit (Local Government Management Board, 1991). It has been accepted, particularly in the oil and petrochemical industries, that production processes can go wrong and employees are occasionally injured. It therefore seemed natural for organizations in these industries to extend their sphere of concern from the local community to the local environment. While the safety audit may have been a logical place to begin, it was very much ineffectual as an environmental strategy because practices and standards varied from country to country (Wheeler, 1993).

Total quality environmental management

Another approach to environmental management was to treat the environmental impact of business activities as a 'quality assurance issue' and two environmental management standards, Total Quality Environmental Management (TQEM) and BS 7750 (BSI, 1994), were introduced. However, these standards of environmental quality measurement were seen as extensions of the existing total quality management (TQM) standards, BS 5750 and ISO 9000, and carried low priority status particularly against daily operational requirements (Tinsley and Pillai, 2006).

Despite the slow uptake by organizations of EMSs, the steady increase in the amount of environmental legislation and the increasing cost of environmental reporting, administration and compliance emerged as two key drivers in moving organizations beyond merely pollution control and to see future environmental investment as a method of reducing compliance costs and minimizing environmental penalties.

Organizations with existing TQM systems saw the benefits of decreasing manufacturing costs, faster time to market and increased market share that would follow by introducing EMSs. However, while some organizations were achieving success with environmental management strategies, others found their strategies had to be shelved or abandoned altogether (Shelton, 1994). The main reasons for such actions were lack of support for the EMS by middle managers, lack of commitment by senior management, and inability to communicate the benefits of further environmental investment. Due to the increasing severity of penalties for environmental management failure, large organizations were beginning to adopt EMSs that focused on the 'management' of environmental risk.

Eco-management audit

The third phase was the environmental audit. This led to the creation of an 'Eco-management Audit', frameworks such as BS 7750, ISO 14001 and the EMAS that were designed for national and international standards (British Standards Institution, 1992; Council of the European Communities, 1992; International Chamber of Commerce, 1994). ISO 14001 eventually superseded BS 7750 and is applicable to all organizations worldwide.

The purpose of an environmental audit is to provide a procedure and a process of checks and balances that form part of a circular, continuous programme of improvement for the EMS (Elkington and Hailes, 1987). The benefits suggested were that the audit leads to the identification of risks as well as pinpointing cost saving opportunities. However, due to the pressures brought to bear in complying with current legislation, too many organizations were only using audits to verify and achieve compliance (Welford, 1996).

Environmental management systems

The fourth phase was the introduction of the EMS. Here, emphasis was placed on 'management' to control policy issues, internal resources, purchasing, product or service design, communication and education. Such emphasis was designed to measure management decision-making and environmental consequences and to make environmental management part of daily operational activity.

Although national and international bodies were setting standards for levelling the business playing field, it still remains that environmentalism, for many organizations, is merely about compliance (Avila and Whithead, 1993). This view is shared by Shrivastava and Hart (1994), who state that a 'command and control' style regulation which forces organizations to approach environmental issues in a fragmented fashion, has produced little progress.

The EMS detailed in Figure 1.1 identifies the type of system that would be required to implement an effective EMS into an organization (Welford and Gouldson, 1993). The system based on a TQM structure is designed to ensure that environmental management is up to a 'quality' standard. The reasoning for such strategic configuration is that similarities between 'quality' systems and EMSs provide management with familiar reference points (Welford, 1996). Conflict can

FIGURE 1.1   Environmental management system

occur, however, if both systems are in opposition and resources and management are divided as to the best option to take (Tinsley and Melton, 1997).

Similarly, each system may vary between organizations depending on its structure and the level of environmental commitment by senior management. The more complex the EMS the more organizational forces act against, rather than for, its successful acceptance into an organization (Roome, 1992).

As some organizational change is required with the implementation of any strategic change it is worthwhile noting that the more radical the change the more resistance may be encountered. More resources are required to overcome this resistance, and more uncertainty is introduced into the organization (Roome, 1992). In short, the more intricate the strategy, the greater the organizational barriers that hinder a successful acceptance. With the introduction of any new strategy, there are always organizational forces acting for and against its introduction.

Those organizations with existing TQM systems saw the benefits of decreasing manufacturing costs, faster time to market and increased market share that would follow by introducing EMSs (Shelton, 1994). However, Shelton's (1994) study found that, while some organizations were achieving success with environmental management strategies, others found their strategies had to be shelved or abandoned altogether. The main reasons he found for organizations 'Hitting the Green Wall' were the lack of support for the environmental management strategy concept, the lack of environmental strategy focus, and the inability to communicate the benefits of further environmental investment.

The existing dilemma for senior management is that they are advised that good environmental management is rewarded with reduced operational costs, improved value of a product, operational efficiency, new marketing outlets, enhanced corporate image and new business opportunities (Taylor, 1992; Hui *et al.*, 2001). They are also advised that environmental management could be the next strategic management tool for competitive advantage (Greeno and Robinson, 1992). This advice, however, is contrary to management's awareness of the business reality that an EMS is just another, newly emerging corporate strategy that must be given the time and resources to be accepted, as any other new strategy, as part of the existing corporate culture (Rothenberg *et al.*, 1992).

Environmental academics and practitioners are aware that organizations are spread along an environmental continuum with 'beginners' and 'proactivists' at opposing poles (Hunt and Auster, 1990). Some of the main reasons for the varying levels of environmental commitment are stated as availability of resources and the commitment of senior management (Greeno and Robinson, 1992; McGrew, 1990). However, the commitment of resources and senior management according to Roome (1992) does not, in itself, ensure the successful introduction of an EMS. An effective EMS, he adds, must promote a quality and commitment, by an organization and its employees, to an environmental ethic.

This history of the development of EMSs provides an insight into the road that has been travelled and the experiences learned along the way by others. The majority of large organizations have EMSs and environmental policy statements declaring that their objective is to improve environmental management and reduce exposure to environmental risk.

Environmental management standards

To address the issue of environmental performance in industry, environmental standards were considered to be the best way to demonstrate the benefits of better environmental management to industry and to offer a process, similar to the familiar BS 5750 quality assurance standard, to reduce the risk of exposure to the increasing numbers of European environmental directives.

BS 7750

BS 7750 was the first UK national standard created for an EMS in the 1990s (Morrow and Rondinelli, 2002). Based on the BS 5750 quality system, BS 7750 was used to describe the organization's EMS, evaluate its performance and to define policy, practices, objectives and targets; and it provides a catalyst for continuous improvement.

The concept is similar to that used in quality standards BS 5750 and superseding ISO 9000, in that the methods are open to definition by the organization. The standard provides the framework for development and assessment of, BS 7750 EMS. BS 7750 was developed as a response to concerns about environmental risks and damage (both real and potential). Compliance with the standard is voluntary and complements the requirements for compliance to statutory legislation. As BS 5750 was the driver for ISO 9001, so BS 7750 led to the development of ISO 14001. BS 5750 and ISO 9001 do not relate to quality in environmental terms but there are many similarities both in terms of management implications and also in the registration process (Ball, 2002).

As its base, BS 7750 requires the environmental policy to be fully supported by senior management policies and to be outlined to both staff and the general public. The policy needs to clarify compliance with environmental legislation that may affect the organization and stress a commitment to continuous improvement. Emphasis has been placed on policy as this provides the direction for the remainder of the management system.

The preparatory review and definition of the organization's environmental effects is not part of a BS 7750 assessment but examination of these data provides an external auditor with a wealth of information on the methods adopted by the organization. The preparatory review itself should be comprehensive in its consideration of input processes and output at the site. It should also be designed to identify all relevant environmental aspects that may arise from the organization's activities. These may relate to current or future operations, as well as to the activities performed on-site in the past (i.e. land contamination).

EMAS

The EMAS is similar in structure to ISO 14001 and was launched in 1995. There are two major differences between EMAS and ISO 14001:

- The whole organization can be certified to ISO 14001 whereas EMAS is generally a site-based registration system.
- Whereas any organization from any business sector can use ISO 14001, EMAS is only available to those organizations operating in the industrial sector.

Within the UK, an extension to the EMAS scheme has been agreed for local government operations, who may also register their EMSs to the EMAS regulations.

In addition to a summary of the process, the statement requires quantifiable data on current emissions from the site and environmental effects, the amount and types of waste generated, raw materials utilized, energy and water resources consumed, and any other environmental aspect that may relate to operations on the site.

Pre-assessment is as much part of the EMAS as it is of ISO 14001. The environmental audit must be comprehensive in consideration of input processes and output at the site. The procedure is designed to enable identification of all relevant environmental aspects that may arise from the site itself. The pre-assessment also includes a wide-ranging consideration of the legislation that may affect the site, whether it is being complied with currently and, perhaps, even whether copies of the legislation are available. Many of the environmental assessments undertaken already have highlighted the fact that organizations are unaware of all the environmental legislation that affects them, and therefore fail to meet the requirements of that legislation.

Under the EMAS standard an organization declares its primary environmental objectives as those that have the greatest environmental impact. In order to gain most benefit, these objectives become the primary areas of consideration within both the improvement process and the organization's environmental programme. The programme is the plan used to achieve specific goals or targets and describe the real and achievable means to be used to fulfil those objectives.

As with ISO 14001, the EMAS standard requires a planned, comprehensive and periodic series of audits of the EMS to ensure that it is effective in operation, is meeting specified goals, and continues to perform in accordance with relevant regulations and standards. The audits are designed to provide operational information in order to exercise effective management of the system, providing information on practices that differ from the current procedures or offer an opportunity for improvement. Under the EMAS the minimum frequency for an audit is every three years.

Most organizations are used to producing an annual report and accounts describing the activities of the organization over the previous year, and their plans for the future. The EMAS generally requires a similar system for the organization's environmental performance. It is also a requirement that there should be notice of any particular plans for the future that may have an effect upon the environmental performance of the organization, whether detrimental or beneficial.

The peculiarity with the EMAS is that the policy statement, programme, management system and audit cycles are reviewed and validated by an external, accredited organization. In addition to providing a registration service, this organization is also required to confirm, and perhaps even sign the organization's periodic environmental statements.

Environmental management standard: ISO 14001

At the same time as the European Commission put forward a proposal for the EMAS, the BSI in the UK also devised an environmental assessment standard called BS 7750. It was also based on the quality standard BS 5750 and was launched after consultation with industry and a two-year pilot programme. The objective of BS 7750 was to put the emphasis on the management of environmental systems, not just the system itself.

The BS 7750 and EMAS standards were very similar in environmental requirements for businesses. At the time it was thought appropriate to make the EMAS a mandatory standard for businesses. Strong industry lobby groups argued successfully, however, that a mandatory approach would be detrimental to industry and the EMAS became a voluntary scheme.

In 1993 it was felt that an international standard was required for environmental management. In 1996, from an idea based on BS 7750, the ISO 14001 standard was born (Morrow and Rondinelli, 2002). It was argued to be: a) a step forward given the successes of the quality standard ISO 9000; b) the standard that would replace the numerous and often conflicting sets of criteria found in various countries and c) it focused on the processes involved in the creation, management, and elimination of pollution rather than just directly reducing pollution (Melnyk et al., 2003).

As BS 5750 had been withdrawn with the appearance of the ISO 9000 series, the outcome of the emergence of the international standard ISO 14001 meant that the national standard BS 7750 together with national standards in other European

Union (EU) countries were, with common consent, also withdrawn. Since the introduction of ISO 14001 many other ISO 14000 standards have also come into operation; the list is shown in Table 1.2.

ISO 14001: the global green standard

The ISO 14001 standard was introduced in 1994, benefiting from the success enjoyed by the ISO 9000 family of standards. ISO 14001 was based on the model of various national environmental management standards, and in particular on the British BS 7750 standard.

The standard was created in the institutional context of the 1990s, a period characterized by the strengthening of the trend towards seeking a green paradigm for production and consumption.

Motivated by different stakeholders and the internal improvement of their general and environmental efficiency, an increasing number of organizations began to introduce an EMS at this time. An EMS is a systematic process that corporations and other organizations use in order to implement environmental goals, policies and responsibilities, as well as to provide for regular auditing of these aspects (Cascio, 1996).

According to Haufler (1999), some organizations introduced EMSs to avoid the introduction of more far-reaching public environmental regulations, while others introduced them to respond to the criticisms made by environmental activists. Others adopted them to enhance the ISO 14001 standard as a process, not a performance standard that describes a system to help an organization achieve its own environmental objectives. It is assumed that by helping an organization focus

TABLE 1.2 ISO 14000 standards

ISO 14001 Environmental management systems – Requirements with guidance for use

ISO 14004 Environmental management systems – General guidelines on principles, systems and support techniques

ISO 14015 Environmental assessment of sites and organizations

ISO 14020 series (14020 to 14025) Environmental labels and declarations

ISO 14030 Post production environmental assessment

ISO 14031 Environmental performance evaluation – Guidelines

ISO 14040 series (14040 to 14049), Life Cycle Assessment (LCA), pre-production planning and environment goal setting

ISO 14050 Standard terms and definitions

ISO 14062 Guidance to improving environmental impact goals

ISO 14063 Environmental communication – Guidelines and examples

ISO 14064 Measuring, quantifying, and reducing Greenhouse Gas emissions

ISO 14065 Accreditation requirements for organizations undertaking Greenhouse Gas (GHG) validation or verification

ISO 19011 Audit protocol for both integrated 14000 and 9000 standards on its manufacturing process, it improves its environmental performance (Melnyk *et al.*, 2003).

BS 8555/Acorn Project

The most recent development designed to assist the smallest to the largest businesses achieve ISO 14001 is the phased EMS implementation approach used by BS 8555 and piloted through Project Acorn. This method focuses on introducing greater flexibility to achieve the ISO 14001 standard by breaking the process down into bite-size pieces. It provides a six-level staged approach which is as follows:

Step 1: A commitment to the ISO 14001 standard and establishing a baseline on which to monitor continuous improvement.
Step 2: Compliance with customer needs and legal and regulatory requirements.
Step 3: Identification of significant environmental aspects and impacts for the development of objectives and targets and the management programme.
Step 4: Management of significant environmental aspects using the management programme.
Step 5: Documentation check, audit and review of EMS.
Step 6: External communication to gain full accreditation against an internationally accepted EMS standard.

After each phase of the scheme has been implemented, the organization can either assess themselves through internal audits, allow major customers to assess them or be assessed by a third party, to ensure that the requirements of each phase have been met (The Acorn Trust, 2005). Undergoing an external audit may have added benefits to both customers and suppliers, as evidence of these external audits may be used to avoid second party audits of their supply chain (Gascoigne, 2002).

In some cases, organizations may not necessarily have to achieve full certification before becoming part of a major supply chain. Achieving phase three of BS 8555 is often enough to be accepted as a supplier. However, more and more organizations are looking for international accreditation to acknowledge their environmental commitment (Khanna and Anton, 2002).

In recent years ISO 14001 certification has been experiencing major growth on the international stage. It is worth noting that by the end of 1999, 14,106 certificates had been issued worldwide, and by the end of 2007 the number of certificates issued had reached 154,572. In the space of eight years, the number of certificates issued worldwide had increased nearly eleven-fold. Attention should be drawn to the fact that around 40 per cent of ISO 14001 certificates issued worldwide were issued within the EU. The USA's share, on the other hand, was limited to 3.5 per cent, while China and Japan were undeniably the world leaders in terms of the absolute number of certificates issued, accounting for 20 per cent and 18 per cent of the total respectively. In recent years, special mention should be made of the growth in the number of certificates issued in the People's Republic of China, due – among other factors – to the influence of the pressure exerted by the Chinese Government to encourage the implementing and certifying of products in accordance with this international standard (Shin, 2005).

Within the European total of certification adoption (Table 1.3) recognition should be given to the increased performance of some of the East European countries that have recently joined the Union. These countries have attracted considerable investment in industrial production and have been experiencing growth levels above the average for European countries in terms of the number of ISO 14001 certificates issued (Heras *et al.*, 2008).

The growth in the numbers of certificates when broken down according to world region suggests a roughly similar pattern of geographical distribution for ISO 9000 and ISO 14000 (ISO, 2001b). Looking at different regions of the world, Europe leads the way in the adoption of both standards with East Asia second and North America third. Many countries in Western and Southern Asia and especially in Africa, with the notable exception of South Africa, still lag behind in certification numbers. This pattern is clearly visible in Tables 1.1 and 1.3, where the number of ISO certificates awarded relates to labour forces in each country in order to allow for the effects of differences in size. In the case of ISO 14001, national differences seem to be, at least partly, attributable to the regulative pressures exerted by the respective environmental policies of each country (Braun, 2005).

EMSs implementation is discussed in greater detail in Chapter 2 including an overview of the main stages, such as audits and aspects and impacts analysis, as well as some practical templates that may be used to structure individual procedures.

Organizational drivers for an environmental management system

Senior managers, introducing an EMS to meet an organization's environmental needs, should have an understanding of what those needs are. Within modern customer focused organizations these days, the drivers for environmental performance and efficiency should be so significant the commitment to environmental

TABLE 1.3 International adoption of ISO 14001 certification

Economic zones	1999	2003	2007	2010	2011
Worldwide Total	13,994	64,996	154,572	251,548	267,457
Africa	129	626	1,096	1,675	1,740
Central/South America	309	1,691	4,260	6,999	7,067
North America	975	5,233	7,267	6,302	7,465
Europe	7,253	30,918	65,097	103,126	106,700
East Asia and Pacific	5,120	25,151	72,350	126,551	137,335
Central and South Asia	114	927	2,926	4,380	4,725
Middle East	94	450	1,576	2,515	2,425

Source: ISO Survey 2011

management should not be on the basis of 'what does it cost to do it?' but rather on the basis of 'what would it cost not to do it?' in terms of protecting corporate reputation, minimizing corporate risk and attracting and retaining staff and investors (Holtom, 2010). However, not all senior managers understand an organization's environmental drivers; if they did there would be greater commitment to developing an EMS and a commitment to the identification of the elements to be included in an EMS to ensure its success (Kirkland and Thompson, 1999). While environmental management improves environmental performance, there are many other benefits depending on the proactivity of the organization and its willingness to increase its capacity in learning, innovation and environmental integration. The various ways organizations commit to environmental improvements suggest that there are different motivations for wanting environmental improvements (González-Benito and González-Benito, 2005).

More specific organizational drivers such as laws (regulations), lawsuits, government policies, banks, investors, accounting systems, employees, markets, costs, public and environmental non-governmental organizations, industry codes and standards, self-regulation, and international factors have been identified by Kirkland and Thompson (1999) and can be incorporated under the main headings in Figure 1.2. The pursuit of these drivers varies from organization to organization, to the extent that some organizations do not recognize the need for any environmental management (Whalley and Whitehead, 1994). Each of the seven key drivers are discussed here separately.

Energy efficiency

This is the most logical starting point for those organizations that wish to begin with something familiar that provides a short-term return for minimum expenditure (Box 1.3). A simple review of oil, electricity and gas bills provides a base from which to measure future savings. Adopting an energy efficiency programme is a good way to begin an environmental awareness programme for the organization. The corporate environmental policy would state that the organization is committed to using energy more efficiently. One of the key objectives might be to reduce

Key organizational drivers
Energy efficiency
Waste minimization
Green organization image
Competitive advantage
Supply chain pressures
Environmental legislative protection
Staff morale and corporate social responsibility

FIGURE 1.2 Key organizational drivers

Box 1.3   Energy efficient investment strategy

Local authorities and government agencies often provide free or low-cost energy efficiency surveys for companies operating within their area. Using a mix of investment strategies can help alleviate available resources issues and achieve significant energy and waste savings.

No investment

A simple programme of energy savings identification could begin with an examination of current electricity, gas, oil, water and effluent bills to ensure that the best possible prices are being achieved from suppliers. A no investment strategy can realize significant savings by renegotiating rates with current energy suppliers or by changing supplier. This strategy identifies immediate financial savings and informs management of the benefits of monitoring energy bills, even if it has no direct environmental benefit.

Low investment

A low investment strategy may involve the purchase of buildings management systems that monitor and optimize energy usage according to room, buildings and workshop occupancy needs and reacting to external weather and seasonal conditions. Additional low investment changes would include switching to energy efficient lighting systems and more efficient electrical appliances.

High investment

Investing in more efficient plant and equipment can involve significant capital expenditure and as a result can constitute a high investment strategy. New technology for more efficient production and manufacturing processes will be more efficient and if planned correctly the initial capital expenditure will ensure future energy efficiency, new products and services for new low carbon markets. They will also conform to the latest environmental legislative requirements.

energy consumption by 10 per cent in the first year measured against existing energy bills.

Waste minimization

Many organizations expend much of their business development budget on improving production or increasing sales. A greater return on investment, as high as 10 per cent of turnover, can be achieved if the same importance or investment is attached to improving waste management. Reducing waste improves profitability and any savings go straight to the bottom line and improve competitiveness. A waste minimization programme improves business efficiency and reduces environmental impact in the short term. Simple product or process design changes can result in fewer natural resources being used to convert raw materials used in the final product. Waste should be seen as product failure (Castledine, 2001; Ackroyd *et al.*, 2003).

Green organization image

Businesses strive continually to be different from their competitors and, in an attempt to gain a competitive advantage, products or services are often linked to environmental benefits (Box 1.4). Many industry sectors are becoming increasingly aware that businesses and the general public prefer, where possible, to deal with organizations that are able to demonstrate a willingness to operate in an environmentally responsible way (Nuki, 2007). Take for example Patagonia, a winter and outdoor clothing company that recognized the benefits to an organization if it was perceived to be protecting the environment. Patagonia's green image went further, when it pledged 1 per cent of its sales to projects that help save the environment. It went further again, on occasion asking its customers not to buy its products unless they really needed to. Organizations willing to take environmentally sensitive stances are seen by customers and staff as having sound ethics and perceived to be honest and forthright in all business matters.

Competitive advantage

If an organization improves its efficiency in its use of resources, particularly in its production processes and use of energy and water, it gains a competitive advantage

---

Box 1.4   Green organization image – General Electric

In a bid to emphasize its green organization image, General Electric (GE) launched, in 2005, its 'Ecomagination' initiative: a company-wide commitment to address global environmental challenges. The initiative is to encourage all GE staff to imagine and develop innovative environmental solutions that benefit its customers and society. This initiative covers the GE product range including energy efficiency in everything from light bulbs to washing machines, locomotives and gas turbines. The 'Ecomagination' initiative is viewed by all GE staff as a business strategy to improve energy efficiency, reduce carbon emissions and to drive growth. To date, revenues earned from the development of environmental technology within this initiative have **exceeded $12 billion annually**.

Bekefi et al. *(2008)*

Box 1.5   Competitive advantage – Manchester United

The football club Manchester United has become the first major stadium in the UK to achieve the international EMS standard ISO 14001. In 2011 the football club committed to reducing its carbon emissions, achieved the Carbon Trust Standard and finished top of the energy efficiency scheme league table. The club has implemented a range of environmental measures including reducing lighting, introducing new controls on heating and air conditioning systems, and launching a communications policy to encourage fans to recycle and travel to matches on public transport. Since its commitment to reducing the environmental impact of its activities, the club has also realized cost and process efficiency savings from waste management. Manchester United football club is seeking further accolades for energy efficiency, waste reduction and low carbon activity as it seeks to be the first major stadium in the UK to achieve the international standard for Sustainable Events Management (Business Green, 2012).

---

over competitors that remain inefficient (González-Benito and González-Benito, 2005). Internally, efficient heating and lighting systems and safe handling of hazardous substances result in greater profitability, improved working conditions and a boost to staff morale which may all, in a smaller, less direct way, contribute to competitive advantage (Box 1.5). Despite the fact that implementing an EMS programme is voluntary, using it as a means of pleasing customers is increasingly more common (Hui *et al.*, 2001). While competitive advantage is one of the more elusive benefits to an EMS (Morrow and Rondinelli, 2002), some organizations actively seek environmental innovations and a competitive priority and these elusive competitive advantages start to become more evident as more organizations require ISO 14001 commitment from their suppliers (Cochin, 1998). An example of this is part of Toyota's strategy to substantially reduce its carbon footprint (Box 1.6).

Supply chain pressure

A study was undertaken by the Business for Social Responsibility Education Fund in 2001 to look at the growing pressure from supply chains to form environmental management strategies. Interviews with 25 suppliers suggested that a growing number of organizations were seeking to address environmental issues across their supply chains. Distributors and dealers in the automotive sector faced the most requests for evidence of environmental strategy from manufacturers (González-Benito and González-Benito 2005). In addition, a number of suppliers had involved their customers in changing product specifications that had precluded pollution prevention activities.

Box 1.6 Green purchasing guidelines – Toyota

In March 2006, Toyota reviewed and revised its environmental purchasing guidelines, which required its suppliers to proactively promote environmental initiatives. These revisions included:

- To implement environmental initiatives together with social aspects of supplier business activities.
- Initiatives that were begun after the initial purchasing guidelines had been issued (such as responses to the EU ELV1 Directive, responses to Eco-VAS 2 and environmental initiatives during logistics activities of contracted transportation organization) have been included.
- Against the background of Toyota's globally expanding environmental initiatives, suppliers are asked to implement environmental measures like CO_2 emissions reduction, in their production activities.
- To further reduce CO_2 emissions during logistics operations, suppliers are asked to implement environmental initiatives in their purchasing and logistics activities.

The recent revision also expanded the scope of supplier categories targeted. Approximately 550 suppliers of equipment, and construction and logistics services were added to the existing list of parts and materials suppliers, increasing the total number of organizations covered by the new guidelines to about 1,000. Toyota plans to gradually expand the application of the revised guidelines to the newly included suppliers through consolidated organizations in Japan and other countries. Using these measures and other approaches, the organization plans to substantially reduce its carbon footprint.

Pressure along the supply chain may come in the form of suppliers having to have a fully accredited EMS or a simpler, yet still comprehensive, questionnaire requiring details of suppliers' environmental practices and performance (Ramus, 2002). Competitive advantage and supply chain pressure go hand-in-hand as customers demand that their suppliers meet their own environmental standard or lose their business, and the competitive advantages of 'greening' an organization becomes more evident as more organizations require ISO 14001 from their suppliers (Cochin, 1998).

Environmental legislation

While managers may not pursue environmentally sound strategies using their own reasoning, they are being forced to by regulations (Azzone et al., 1997). Regulatory requirements are acknowledged as being the most important determinant of the number of staff an organization commits to environmental or health and safety issues (Khanna and Anton, 2002). The penalties for organizational transgressions have grown to incorporate significant fines and imprisonment for its directors. Those organizations that have polluted in the past can no longer escape their clean-up responsibilities. The 'polluter pays' principle is central to existing environmental legislation and ensures that pollution caused by organizations in the past is still their responsibility today (Box 1.7).

The majority of large organizations' environmental policy statements declare that the objective is to improve environmental management and reduce its exposure to environmental risk (British Standards Institution, 2000). The penalties for transgressing environmental legislative demands are high and are quickly becoming all encompassing (Ball and Bell, 1997). Despite these growing environmental pressures, however, most managers still hold to the notion that pollution pays but pollution prevention does not (Whalley and Whitehead, 1994).

Environmental charters

A charter is a formal statement of the rights of an organization or a particular social group, which is agreed by or demanded from a ruling or government. In the case of environmental charters, there are many organizations, industries, or governments that show their commitment to environmental issues by signing up to a charter.

Box 1.7   Environmental penalties

Honeywell

The United States (US) Department of Justice (DOJ) announced that Honeywell Resins and Chemicals LLC (Honeywell) has agreed to pay a US$3 million civil penalty for alleged Clean Air Act violations at its Virginia plant, as well as to improve the facility's air pollution control equipment and processes. Honeywell, 'the world's largest single-site producer of caprolactam used in the production of nylon, and ammonium sulphate used for fertilizer', was charged with violating limits on 'emissions of nitrogen oxide (NOx), benzene and other volatile organic compounds (VOCs) and particulate matter'.

Honeywell was also charged for alleged non-compliance with 'requirements to upgrade air pollution control equipment, to detect and repair leaks of hazardous air pollutants, and to develop safeguards on benzene waste'.

Honeywell will also undertake a US$1 million mitigation project at its facility. Under the settlement obligations, Honeywell is expected to reduce its annual emissions of NOx by around 6,260 tonnes, and cut annual emissions of benzene, other VOCs and hazardous air pollutants by 100 tonnes.

BP

On March 11, 2011, the US DOJ formed the 'Deepwater Horizon Task Force' to consolidate several federal agencies' investigations into possible criminal charges stemming from the explosion and spill. On November 14, 2012, the DOJ announced that BP and the DOJ had reached a $4 billion settlement of all federal criminal charges related to the explosion and spill, the largest of its kind in US history. Under the settlement, BP agreed to plead guilty to 11 felony counts of manslaughter, two misdemeanours, and a felony count of lying to Congress and agreed to four years of government monitoring of its safety practices and ethics. BP also paid $525 million to settle civil charges by the Securities and Exchange Commission that it misled investors about the flow rate of oil from the well. As part of the announcement of the settlement, BP said it was increasing its reserve for a trust fund to pay costs and claims related to the spill to about $42 billion. On the same day, the US government filed criminal charges against three BP employees; two site managers were charged with manslaughter and negligence, and one former vice-president with obstruction. Near the end of November 2012, the US Government temporarily banned BP from bidding on any new federal contracts, citing the company's 'lack of business integrity'. As of February 2013, criminal and civil settlements and payments to the trust fund had cost the company $42.2 billion.

Source: United States Department of Justice (March 2005)

Staff morale

One major indirect benefit of implementing an EMS is improved staff morale. While this is unlikely to be the primary driver for implementing EMS in all but the most socially conscious businesses (González–Benito and González–Benito, 2005), it is becoming a greater priority with the advent of 'Corporate Social Responsibility'. In the same way that a greener image can create greater customer satisfaction and loyalty, commitment to the environment can add to job satisfaction and staff loyalty (Box 1.8).

Box 1.8   Greenhouse Challenge

Greenhouse Challenge Plus is part of the Australian Government's comprehensive Climate Change Strategy, announced in 2004. The programme is managed by the Australian Greenhouse Office as part of the federal Department of the Environment and Heritage.

The aims of the Greenhouse Challenge Plus initiative are to:

- Reduce greenhouse gas emissions.
- Accelerate the uptake of energy efficiency.

- Integrate greenhouse issues into business decision-making.
- Provide more consistent reporting of GHG emissions levels.

As part of the initiative, Greenhouse Challenge Plus has actively encouraged businesses to educate and motivate employees so that they change their workplace habits on a day-to-day work basis. The Department of the Environment and Heritage states that one of the important indirect benefits of staff involvement in the initiative is that it boosts staff morale, which in turn contributes to higher productivity. They also acknowledge that the staff may have expertise that could contribute to a continuous improvement of greenhouse performance in the future and the benefits of creating environmental 'champions' within the workplace, who will motivate others (Australian Department of the Environment and Heritage, 2005).

Interventions for drivers

In an effort to understand why organizations move towards environmental improvements a study (Khanna and Anton, 2002) compared two scenarios: the threat of regulatory penalties or the realization that environmental improvement offered opportunities for business improvement. The study also looked at the types of intervention that would result depending on whether the perception was one of threat or opportunity. These interventions were categorized into two types. A Type I intervention included internal management procedures such as policy development, audits, operations and procedures, setting corporate standards, contingency funding for liability costs, insurance and appointment of environmental staff. A Type II intervention included activities external to the operations of the business such as understanding the supply chain in greater depth and engaging suppliers, their staff, stakeholders and customers, through training, publishing of environmental reports and producing environmental policy statements, to improve environmental performance.

The study concluded that management faced with threats such as legislative and regulatory requirements, audits and inspections led to the adoption of Type I interventions, with little likelihood of adopting Type II interventions. Opportunities such as closer contact with customers, greater chance of innovation and added value, led to the adoption of both Type I and Type II interventions.

Progressing environmental management for a low carbon economy

Public awareness of the impact of business activities on the environment has grown considerably over the past 30 years. While the main objective of economic policy is to maintain and gradually improve a standard of living for the existing population and for future generations, the main objective of environmental policy is to avoid or limit the problems with pollution, dereliction, and the loss of habitats and wildlife species arising from business and human activity. Traditionally, conflict has arisen when the cost of avoiding environmental damage is perceived as being a constraint on economic activity, but with issues such as resource depletion and pollution of the natural environment becoming more prevalent. In an attempt to minimize environmental impact and public attention, organizations have recognized the importance of environmental management and adopted essential environmental standards and systems into daily business activities. An increasing number of organizations are seeing the advantages of environmental business opportunities by these bodies progressing programmes of waste reduction, material reuse and recycling, and energy efficiency by introducing EMSs. In addition, it is recognized that an efficient EMS can influence an organization's structure, strategy, responsibilities, practices and procedures.

The ISO has also given due attention to the question of integrating ISO 9000 Quality management systems (QMS) and ISO 14000 EMS to create value. According to Zeng *et al.* (2005), the similarity and compatibility of the two standards are the main reasons for pursuing such integration. Empirical studies of such integration have suggested that organizations can benefit from avoiding the duplication of procedures, reducing conflict among procedures, and reducing their requirements for resources (Zeng *et al.*, 2005). The basic differences between ISO 9000 and ISO 14000 relate to the policies and objectives of the two systems (Affisco *et al.*, 1997), with ISO 14000 being significantly more committed to stakeholder satisfaction and improving relationships with stakeholders (Poksinska *et al.*, 2003).

In recent developments, revisions of the ISO 14000 standards have been released in the form of ISO 14001: 2004 and ISO 14004: 2004. The revisions have brought ISO 14001 more closely into alignment with the ISO 9000:2000 QMS standards. Nevertheless, general criticisms of the ISO 14001 standard continue – including the fact that the system lacks a requirement for public reporting and a benchmark during implementation, which is believed to be the impending fourth standard for social responsibility (SR). ISO 26001 could solve the limitations of ISO 14001.

The term sustainable development (SD) was defined in the Brundtland report (1987), and later extended by the World Business Council on Sustainable Development (WBCSD, 1996), to describe a new conception of corporations with respect to what they are, what they do, and how they relate to social, environmental, and political issues (Demirag *et al.*, 2005). According to these authors of the report, SD provides an opportunity for a variety of stakeholders and interested persons to meet and use a common language when discussing solutions to shared problems.

In discussing the significance of SD, Hart (1997) stated that the global nature of contemporary political and social issues now exceeds the resources, technology, and reach of any single organization. To achieve SD in these circumstances, it is necessary for organizations to adopt a policy of corporate social responsibility (CSR) (Zadek, 2001). According to Edvardsson *et al.* (2006), this requires the adoption of a philosophy of 'values-based value creation', whereby the economic, social, and environmental value-creating processes of an organization are integrated on the basis of shared values and shared meanings.

Freeman's (1994) 'stakeholder theory' constituted a major theoretical advance in incorporating ethical considerations into organizational design. According to Laszlo (2003), the transformation of organizational culture to include consideration of the legitimate interests of key stakeholders has enabled the concept of sustainability to become an essential aspect of business conduct.

Pruzan (1998) also adopted elements of 'stakeholder theory' in talking about a contemporary shift in perspective from a focus on 'control' to a focus on 'values'. According to the author, this calls for a shift to 'values-based management', which is premised on a stakeholder perspective of leadership, responsibility, and ethics. In a similar vein, Vargo and Lusch (2004) talked about a contemporary shift from the traditional 'logic of goods' to a new dominant 'logic of service'. Taking a lead from these paradigm shifts suggested by Pruzan (1998) and Vargo and Lusch (2004), the concept of 'values-based value creation' was developed by Edvardsson *et al.* (2006). This concept also adopts a stakeholder perspective in placing the values inherent in an organization's core strategy and culture at the centre of a web of stakeholder relationships to create value for owners, customers, employees, and society at large. It is argued that such a transformation leads to the adoption of CSR and sustainability practices that extend to employees, business stakeholders, and even competitors (Laszlo, 2003).

In developing the concept of 'values-based value creation' described above, Enquist and Edvardsson (2006) also indicated that CSR and SD can act as driving forces for value creation through enhanced service quality. As Vargo and Lusch (2004) are asserting that a shift is taking place from the traditional 'logic of goods' to a new dominant 'logic of service', the quest for improved service quality must be an indispensable business strategy, as has also been asserted by Kim *et al.* (2004). In seeking such an improvement in service quality, the question of how customers assess 'value' in goods and services is obviously central. In this regard, there is a growing realization that customers do not value physical goods as such but see tangible products as 'platforms' for service experiences and experience-based quality (Edvardsson, 2005). Service quality, therefore, is perceived and determined by the customer on the basis of experiences, activities, and interactions that lead to solutions to that customer's problems.

In this context, Schneider and White's (2004) suggestion that TQM should be integrated with service offerings in operations management provides evidence that contemporary business trends are in alignment with an overarching view of the importance of sustainability. The integration of different internal management systems with a view to enhancing the delivery of quality service signifies new ways of thinking about sustainability, values-based value creation, and service quality.

An EMS has been shown to have a positive impact on overall organization performance (PricewaterhouseCoopers, 2002; Melnyk *et al.*, 2003). As organizations continue to develop systems to reduce environmental risk, so more are incorporating sustainable practices into this all-encompassing risk management system.

The approach of incorporating sustainable development into an EMS that also includes quality, corporate ethics and occupational health and safety (OHS) management reflects both the origins and the evolution of SD. Environment, Health and Safety (EHS) management systems, developed in response to the need to address adverse consequences of industrial activities, ensure compliance by addressing business issues with an insider's knowledge of operations and provide a sustainable level of protection of employees, the public and the environment.

By contrast, SD is focused on the core of the business itself. It looks to long-term future impacts of the business with respect to the utilization of resources and the provision of goods and services. Once the organization has identified its vision in terms of sustainability the use of one management system is preferred as opposed to creating a system of systems. Having a commitment towards SD would put a greater emphasis on the external impact of an organization's operation, thereby enhancing the positive impact on corporate performance, but also encouraging a more responsible, ethical, environmental and social corporate culture.

Corporate social responsibility

CSR outlines specific corporate characteristics and behaviours and describes how managers manage the interests of corporate constituencies. It reflects and explains past, present, and future states of affairs of corporations and their stakeholders. The instrumental approach establishes a framework for examining the connections between the practice of stakeholder management and the achievement of various corporate policy goals. The normative approach involves the notion that all stakeholders' interests are of intrinsic value; that is, that each stakeholder group merits consideration for its own sake and not because of its ability to further the interests of some other group, such as shareholders. The strategic approach to CSR, while anchored in the justifications made by the earlier schools of thought, offers a framework organizations can use to identify the social issues that benefit stakeholders while simultaneously strengthening the organization's competitiveness.

It has been argued (Porter and Kramer, 2011) that such an approach to CSR results in a symbiotic relationship between an organization and its stakeholders since the success of the organization and the success of the stakeholders become mutually reinforcing. While staff morale is not often directly measured as part of an EMS, anecdotal evidence is available (Morrow and Rondinelli, 2002; Hillary, 2004). Managers are increasingly becoming more aware that stakeholder perceptions can be critical to corporate performance – and sometimes even survival (Boesso and Kumar, 2007; Orij, 2010). CSR initiatives have come under closer scrutiny in terms of the business benefits received from supporting the demands of disparate groups of stakeholders. Given the recent financial crisis, accurate assessment of the benefits received from various CSR initiatives has taken on new importance, as corporate boards try to balance their organizations' social obligations with the pragmatic imperative of the most effective utilization of shrinking resources (Boesso and Michelon, 2010; Maine and Sprinkle, 2010). Therefore, accurate measurement of the benefits received from CSR initiatives takes on added importance from both the organization and the stakeholder perspectives. It is not surprising, then, that the number of organizations and agencies that evaluate and rank organizations on their corporate social performance has increased in recent years.

- The idea of seamless and effective integration of standard principles into corporate governance. Several organizations issued guidelines for implementing SD either as a stand-alone programme or as part of ISO 14001 (GEMI, 2002; Gilding *et al.*, 2002; Timberlake, 2002; Rittenhouse, 2003).
- A complex set of organizational variables, including corporate structure, individuals' skills, corporate culture and resources, play an important role in the success of any structural reorganization around an operational standard (Burke and Litwin, 1992).
- The provision of a framework for future enterprise resource planning begins by identifying key elements, practices and key links business activities where an EMS and SD are fully integrated. The advantages this provides to organizations are many (Buckingham, 2003).

2

ENVIRONMENTAL MANAGEMENT SYSTEM DESIGN AND IMPLEMENTATION

Summary

Utilizing a step-by-step approach, this chapter provides a detailed insight into the practicalities involved in implementing an EMS into an organization. Based on ISO 14001 standard guidelines, a framework of an EMS is outlined and discussed. It includes operational functions such as environmental audits, impact and aspects assessment, keeping abreast of legislative requirements, the resource commitment required and the mechanisms needed to ensure continuous improvement. In addition, various options for customizing the EMS are discussed, including integrating systems such as quality, health and safety, social and corporate responsibility. The chapter concludes with a detailed discussion on the formulation of a corporate environmental plan.

Organizations are becoming more proactive in their approach to dealing with environmental risk associated with their operational activities, motivating actions to integrate environmental issues into business decision-making. As a consequence, formalized sets of environmental management practices have been adopted to satisfy independently certified, EMS, criteria. This chapter discusses the basic framework of an EMS to assist its design and implementation into an organization. It includes a detailed understanding of activities such as audits, impact and aspects assessment, keeping informed of legislative requirements, the resource commitment required and the mechanisms needed to ensure a programme of continuous improvement.

The implementation of an EMS is an exercise that requires the integration of environmental issues into every aspect of an organization's business activities and the engagement of all staff and stakeholders. To achieve quality of products or services, management must have control of every aspect of its business, including environmental and social aspects as well as the economic imperatives (Yin and Schmeidler, 2008).

Control is gained by:

- finding out what needs to be controlled;
- deciding how to control it;
- implementing a system of control; and
- maintaining control.

EMS structure

This chapter provides an outline of a basic EMS based on ISO 14001 international standard and an implementation process to enable an organization to provide a foundation system to manage its environmental risk at the basic compliant level. A basic EMS provides a foundation on which to build or customize an EMS that suits the needs of an organization and the demands of its various stakeholders. Such a system should also reflect an organization's cultural commitment to providing ethical and, renewable sourced resources in the production and delivery of its products and services. A flexible EMS foundation also enables the integration of health and safety, quality and occupational health systems into one coordinated, integrated system. These systems may manifest themselves as Safety, Health and Environment (SHE), Environment, Health and Safety (EHS), TQEM and other such integrated systems. Having an EMS foundation offers an organization the ability to build in its commitment to CSR and SD at national, international and global levels.

Implementing an EMS enables all four aspects of control to be achieved, and also ensures that environmental issues are given due consideration along with finance and product quality. From the research literature an EMS has been found to have a positive impact on an organization's bottom line by providing a mechanism for continued improvement, resource efficiency, customer needs and ultimately financial performance (Griffith and Bhutto, 2008). The impact of environmental activities on corporate performance has been shown to be strongly affected by the presence of a formal EMS. The improvement generated by the presence of a certified, formal EMS can be explained by the way it helps to involve people in the environmental activities of an organization. Alternatively, the evaluation by an impartial third party may, similarly, motivate improvements. This gains the organization long-term improvements, not only in the decreased levels of pollution and waste generated but also by increasing operational performance (Melnyk et al., 2003).

The components of an EMS management plan are summarized in Figure 2.1 and each aspect is discussed in further detail below. An EMS such as ISO 14001 is based on Deming's PDCA (Plan-Do-Check-Act) management cycles (Ammenberg and Sundin, 2005). The ultimate aim of an EMS is to produce a corporate environmental plan which provides organizations with a route map to improved environmental performance. However, as is seen in this chapter, the key to successful environmental management improvement is constant monitoring and measurement of the effectiveness of the EMS and reviewing and updating the corporate environmental plan.

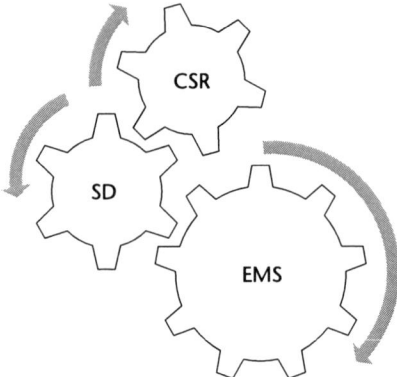

FIGURE 2.1 EMS systems mechanism

Management commitment

To ensure the successful installation of an EMS, it is essential to engage full management commitment. People and budgets must be made available together with the dedication of one or more of the senior management team to the project. If possible, recruit at least one director onto the working group. Lack of support from senior managers makes the implementation process difficult, if not impossible. There are many political issues to overcome throughout the implementation process and more efficient headway is made if a director, or senior manager, is a member of the implementation working group.

Environmental policy statement

An organization wishing to achieve ISO 14001 certification, or similar environmental standard, is required to state its intentions in an environmental policy statement. An organization's environmental policy is, typically, a one-page primary statement by which it communicates its commitment to the management of its business activities and their impact upon the environment. Within this policy statement an organization signals its commitment to realistically achievable objectives and targets together with an aim of continuous environmental improvement. The standard states that a senior manager must be involved during all stages of the development of the policy statement including policy review, updating, approving and final signing.

Within the scope of an organization's activities, products and services, senior management commitment should include a signed environmental policy statement, which states their commitment to protection of the environment, prevention of pollution and continuous improvement of an organization's environmental performance. It forms the framework on which the EMS operates, as well as a signal to internal and external stakeholders that the organization is serious about protecting the environment (Smith and Sharicz, 2011).

There is always scope within an organization to continually reduce its environmental impact regardless of its business activities. The following list presents some examples of areas where abundant opportunities exist for continual environmental improvement:

- Minimize energy use through effective energy management.
- Introduce programmes to minimize waste.
- Reduce emissions.
- Ensure new processes are fully assessed prior to introduction.
- Reduce the likelihood of environmental incidents.

The policy statement should also set out the philosophy and principles of an organization with regard to the achievement of the stated objectives and targets. These policy goals should not be too detailed at this stage as it may result in constant updating and re-issuing of the policy. However, general, unsupported, statements of change are not acceptable as organizational goals. Some examples of useful policy objective and target commitments can be stated as follows:

- Provide adequate resources and personnel to maintain the EMS.
- Ensure the employees are informed and receive adequate training to understand their responsibilities in respect of the environmental policy.
- Ensure that the requirements of environmental legislation are understood and met and, where possible, exceeded.
- Provide a framework for identifying and setting environmental objectives and targets.
- Integrate environmental considerations into the design of products and services in order to avoid or minimize environmental impacts.
- Monitor environmental performance continuously.
- Review and audit the effectiveness of an organization's environmental policy.
- Where it is feasible, work with suppliers, contractors, and subcontractors to improve their environmental performance.

The role played by senior management in the development of the policy is also to be audited and reviewed during the certification process.

Environmental audit

The environmental audit requires an analysis of the current level of environmental risk and where an organization stands in relation to it. The audit is a prelude to determining the future objectives of an organization and the procedures for achieving these objectives. It is a process of extracting information about an organization that, when analysed, provides a realistic assessment of its impact on the environment and identifies a set of environmental objectives and targets to reduce this impact. The establishment of these objectives and targets forms the basis of the corporate environmental plan (see Figure 2.2 and 'Creating a corporate environmental plan' at the end of this chapter).

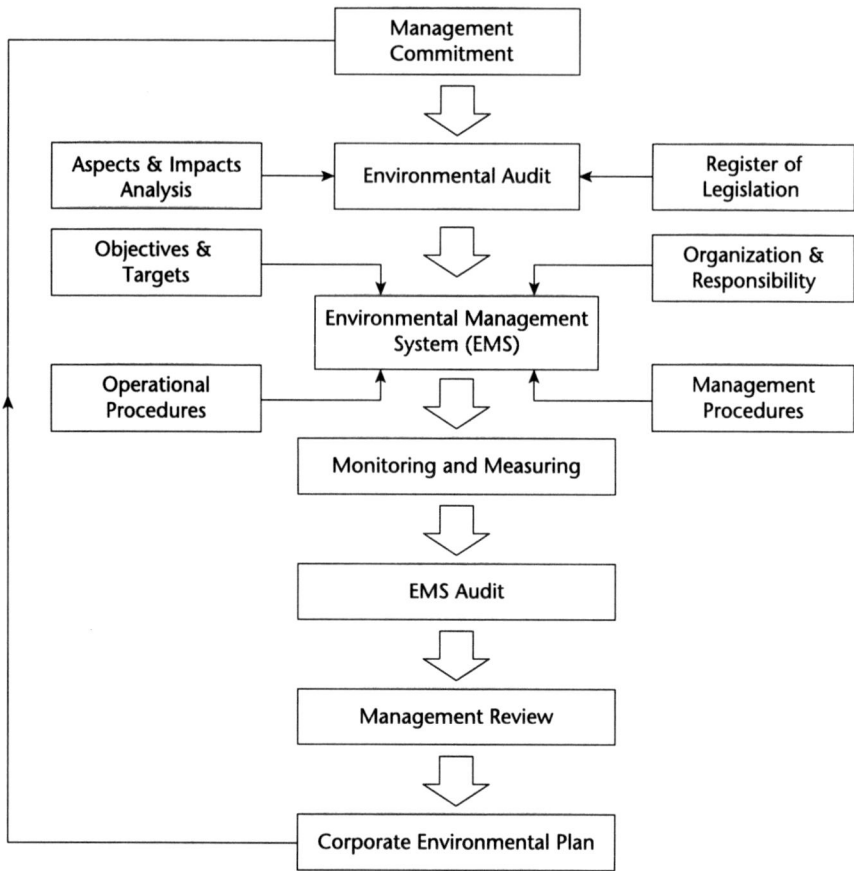

FIGURE 2.2 An environmental management plan

An environmental management audit is often lengthy, with the implications summarized as part of an aspects and impacts analysis. Therefore, an environmental management audit is not usually included as part of the environmental management plan, presentation or report. If it is required in the report, it should be included as an appendix at the end of the plan.

Any environmental management audit includes an analysis of external, as well as internal factors. External factors may arise from the environment, customers and other stakeholders. They relate to circumstances that an organization cannot control directly and include broad factors such as the economic situation, customers, competitors, and political change.

Audit process

The audit process for an EMS has two parts. First, the pre-EMS audit provides knowledge of what the business activities are and what impact they have on the environment (internal audit) as well as those forces outside the business that may not be easily controlled (external audit). Some of these activities can have a positive and negative impact upon the environment. The information gained from an audit provides an evidence base for the chosen environmental management objectives and targets. The second audit, the EMS audit, is designed to ensure that the EMS and its associated procedures are in place and actually work. This second part therefore only occurs once the EMS has been implemented and sufficient time should be given for the EMS and its procedures to function before this final audit is undertaken. Implementing ISO 14001 requires that the EMS operates for a minimum period of three months before final auditing. The term environmental audit is not a clear one, as potentially there are many different types of audit; however, for the purposes of implementing an EMS the following is a list of the most common:

- Internal EMS audit.
- Due diligence audit.
- Environmental review.
- Waste minimization audit.
- External EMS audit (by certification body).

Internal audit

An internal environmental audit is undertaken to assess the environmental impact of an organization's operations and of the products and services it currently produces. From this internal audit a list of operating aspects having an impact upon the environment can be created. An internal audit enables management to determine the level of impact each aspect has and how significant it is to an organization's operations. The results of an internal analysis are used to form the basis of the aspects and impacts analysis that is explained later in this chapter.

Factors in an internal appraisal

Identify the most relevant areas and activities in an organization to appraise. Use discretion and judgement when selecting areas to appraise, as the most relevant internal factors vary between industries, organizations, and individual departments. A useful starting point is to list the major areas of an organization such as departmental activities and resource usage that are critical to financial sustainability. Essentially these are those activities and resources that result in outputs that are valued by the existing customers and it may be useful to seek their views as an aid in the identification process.

A checklist of some of the factors and areas which the internal part of the environmental management audit is given below.

Environmental management

- Environmental research and information systems.
- Environmental planning systems.

- Staff attitudes.
- Organization image.

Finance

While checking the overall finance of the organization as part of the internal audit, it is prudent to establish the likely cost of implementing the EMS, as it varies from organization to organization and across industry sectors. The cost of implementing an EMS should be assessed against the cost of lost revenues should an environmental disaster occur. Critically, identify a dedicated budget and the availability of business resources for the period of the implementation process. Of these available resources establish whether environmental projects are given the same considerations as other business requirements.

Personnel

Taking time to identify the gaps in training, motivation, available skills etc. and establishing development programmes aids swifter implementation. Where appropriate, utilize existing quality, health and safety or similar training structures to which environmental training can be added effectively. This saves on organizational resources and minimizes operational disruption. Determine what levels of environmental management experience and expertise exist within the organization. Similarly, determine the levels of motivation and training requirements for those members of the implementation work group.

Research and development (R & D) and design

When utilizing the R&D and design capability, consider what environmental changes to existing products and services are likely to bring, and what energy efficiency technology and practices can be employed. Consider also, whether some environmental changes allow an organization the opportunity to offer low carbon products to the market, or does a new, or modified, process offer additional materials savings or an additional by-product. Many by-products have emerged from organizations looking at alternative uses for their product or process waste. Identify the level of technological expertise and the role to be played by innovation and design in the EMS.

Engineering and production

Focusing on the engineering and production activities of the organization may identify current and future challenges or present new opportunities. Areas for investigation may include some of the following:

- Technology: materials, components, and machines.
- Techniques: methods, systems.

- Design, promotion, presentation.
- Changes in legislation.

Identify age of plant and its flexibility and cost of any proposed changes when adopting and implementing, for example, new energy efficiency and waste minimization systems. Ensure minimum disruption to supply chains and procurement processes. Closer inspection of existing methods of operation and the identification of alternative uses or changes to the production processes or engineering methods can bring significant benefits to material usage, waste minimization and energy efficiency.

External audit

The external environment encompasses all externalities: the forces and factors outside of an organization that have some impact on its activities but over which it has little, or no, influence or control. The first challenge is to identify the key factors pertinent to a particular department or unit and, where appropriate, the organization. In this phase consider the environmental issues that are current, or have the potential to affect future operations. The biggest issue for most organizations at present is the growth in environmental legislation that has, or should have, an immediate impact on its operational policy and decision-making process. However, in broad terms, it is possible to identify distinct groups of environmental factors, which, to a greater or lesser degree, potentially affect all industries and organizations. These include the influence of environmental pressure groups, which can draw negative publicity to an organization identified as having limited environmental awareness. More importantly, there is also influence from an organization's own stakeholders who are becoming increasingly knowledgeable in environmental, social and ethical matters. Additionally, there are competitors that may have their own strategy for environmental performance, one that gives them a competitive edge. All of these factors should be listed as potential environmental pressures during an external audit.

Organizational stakeholders

Environmental pressure groups

Although these groups may apply pressure in different ways they are all very strong in the public relations and political lobbying arena. These groups are not afraid to take on the largest of organizations and they are adept at coming out on top in media battles. These days, multinational organizations seek to create working partnerships with environmental protection and humanitarian organizations and undertake a period of consultation before any actions are taken. Examples of environmental pressure groups include:

- Friends of the Earth.
- Greenpeace.

- WWF.
- Local and regional 'one issue' pressure groups can also be very important.

Consumers

Environmental issues, and social welfare issues, it is argued, gain greater prominence during periods of economic prosperity, only to return to a lower priority during periods of economic recession. Consumers' concerns therefore may be affected by job security, income levels and confidence in economic conditions.

However, as environmental issues such as climate change and SD are receiving increased media coverage, there is an acceptance that being more responsible for the earth's limited resources is a priority, for the sake of future generations, even if it comes at a higher price. As consumer awareness grows of the environmental damage that can be caused by industry, they are being drawn, mostly by an ethical view, to purchase those products and services that are environmentally friendly, or those that have EMS accreditation such as ISO 14001 (Hui *et al.*, 2001). Banks, financial institutions and various other organization stakeholders are becoming more environmentally aware about where they invest their money.

Competitors and competitive advantage

The implementation of EMS is being accepted as an image-builder and business strategy to strengthen the organization's competitive position and although implementing an EMS is voluntary, using it as a means of pleasing customers is increasingly more common (Hui *et al.*, 2001). As competition for products and services grows so the competitors strive to be uppermost in consumers' minds. The identification of shopping patterns and consumer behaviour is an important tool to winning customers and an awareness of consumer needs. The use of sophisticated software to monitor consumers' buying habits is an attempt to maintain or create an advantage over the competitor. Identifying and anticipating changing lifestyles, such as the move towards organic food by environmentally conscious consumers, adds value to producer's products and services if they are in tune with attitude changes, which can be affected by:

- The changing age structure of the population.
- Trends in family size.
- Changes in the amount and nature of leisure time.
- Changes in attitude towards health and lifestyles.
- Improved education.
- Changes in attitudes towards family roles.
- Changing work patterns.
- Equal opportunities.
- Culture.

It is sometimes a great temptation for organizations, considering the immediate benefits of having a positive environmental image, to cut short the process and virtually change labels on their products and call them environmentally friendly. Such a practice now comes at a higher financial and reputational cost than a change to the product or manufacturing process.

Social media

The internet, mobile phones and social networking are some examples of how an organization may improve its business reach through better communications. New developments in technology such as automation may lead to greater energy efficiency and waste reduction. New environmental legislation may result in harmful products, such as chlorofluorocarbons (CFCs), being banned from the production process. These are just some of the operational policy changes that need to be monitored and assessed for environmental impact. Some of these changes have a positive impact upon the environment and some have a negative impact. The introduction of a green transport plan reduces car usage to and from work, has a positive environmental impact and saves people money, particularly with the ever increasing price of petrol and diesel.

It is not possible to list each and every external environmental development that may result in potential opportunities or threats in the external environmental audit. An attempt to determine the key environmental forces and factors that need to be assessed is sufficient, as well as understanding that these may vary between departments.

Sometimes the most relevant external environmental forces are not immediately obvious as new forces may develop or emerge, and existing factors can change very rapidly. It is therefore prudent to keep a broad perspective on what might constitute significant environmental forces and factors and review environmental plans on a regular basis. It is important to try to forecast both the magnitude and direction of trends and changes in those external environmental factors identified as being most significant. Use as many sources of information as possible to forecast possible changes including secondary data such as that found in government or industry association statistics and trade directories.

The timescale that such forecasts encompass is important and varies greatly depending on the sensitivity of the industry. As a rule of thumb the time horizon for external environmental forecasting should be approximately twice as long as the duration of the corporate environmental plan, and this emerges from implementing the EMS.

The essence of effective environmental management involves achieving a strategic fit between the organizational activities and the environmental threats that exist in the environment. The combination of an internal and external audit should identify key factors, consumer trends and potential changes in these factors, as well as the performance and resources of the organization. The mechanism for moving from the information provided in the environmental management audit to using this information in developing an environmental management plan is provided by the aspects and impacts analysis which is discussed next. Essentially, such an analysis is used to develop a plan that builds on identified strengths and avoids or reduces environmental risk.

System monitoring and measuring

Familiarization with the processes and the monitoring and measurement of activities is a key stage in improving an organization's environmental management performance. From this knowledge, an analysis can be undertaken of the associated aspects and impacts analysis arising from an assessment of the organization's activities and processes and their impact on the environment, examples of which are illustrated in Table 2.1.

Aspects and impacts analysis

The scope of the aspects and impacts analysis is required to cover all business activities, products and services over which an organization has direct control and can influence change. The identification of the environmental impacts arising from the environmental aspects forms the foundation to determine the objectives and targets of the corporate environmental plan.

When setting the scope of the aspects and impacts, examples of the types of activities and processes that may need to be considered are shown in Table 2.1. It should be noted, however, that there are different levels of activity. A department or building may be a starting point but then different levels of activity within the building, for example, paint spraying and finishing may also need to be identified. And within paint spraying a level of hazard environments or wastes may also be identified. In general, environmental aspects of an organization can be identified by examining impacts or emissions to air, land and water.

TABLE 2.1 Activities and processes

Site History and Developments	*Workshop facilities*
Personnel departments	Office facilities
All product processes and areas	Catering facilities
Storage of raw materials and products	Contractor activities
Logistics department	Storage of hazardous materials
Supplies and procurement activities	Site emergency planning
Employee off-site activities	Financial and insurance activities
PR and communications	Works services
Buildings and site management	Management activities

Most direct impacts, those within the control of the organization, must be included in the aspects analysis. Indirect impacts, as a result of the involvement of suppliers for example, may well be within the influence of an organization but not within their direct control.

During the early years of an EMS, it is acceptable for the aspects and impacts analysis process to concentrate on the direct impacts associated with an organization's own activities, rather than to consider the indirect impacts caused by external stakeholders, such as suppliers. If this approach is adopted, it is good practice to clearly state those areas that will be included in the long term and when this will happen.

The completed aspects and impacts analysis sheet provides a record of the environmental effects of operating procedures, incidents, accidents and potential emergency situations that may arise. This assists in the monitoring of their effects on the environment and the development of any necessary remedial actions and procedures. The aspects and impacts analysis provides a positive and negative record of all business activities that are significant environmentally.

The formulation of the aspects and impacts analysis sheet is completed at an early stage of implementation; be aware that the analysis sheet should be updated as and when required. Any new equipment brought into the organization, or any process change requires to be assessed for the degree of environmental impact and its significance tested to see whether it should also be included.

For the sake of consistency and standardization the definitions offered here are those laid down in the ISO 14000 standard.

ISO 14001 guidelines describe an environmental aspect as:

> an element of an organization's activities, products or services that can have a beneficial or adverse impact on the environment . . .
>
> (p. 20)

An environmental impact is described as:

> the change that takes place in the environment as a result of the aspect . . .
>
> (p. 20)

The following list details the stages involved in aspect and impact analysis and is discussed in further detail in the following section:

- Identify and list those areas of an organization's operations that may harm the environment.
- Classify the aspects into levels of environmental impact, i.e. low, medium or high.

Using a scale of 1–5 determine for each aspect the probability of impact occurrence.

- Create the aspects and impacts analysis matrix and calculate the impact scores for each aspect.

- Using a mid score, or a cut-off score, select all those aspects above the mid score to create the register of aspects and impacts.

Register of aspects and impacts

The register of aspects and impacts is a list that details any impacts under the following headings and is shown in Table 2.2.

- Item: A serial number used for recording and sorting purposes.
- Environmental Aspect: Brief details of the organization's activity, product or service that can have an impact on the environment.
- Environmental Impact: Brief description of the interaction or change.
- Cause: Specific details of the activity or process causing the environmental change or impact.
- Effect: Details of whether the environmental interaction or change has an adverse or favourable environmental effect.

Examples of aspects and impacts

Once the operational activities of an organization have been established, list all the activities and processes and divide the list into three key areas. First, identify those activities and processes that affect the environment during normal operating procedures. Second, identify all those activities and processes that affect the environment during abnormal procedures and third, identify occasions when incidents, accidents and other emergencies have affected, or had the potential to affect, the environment.

- Normal operations.
- Abnormal operations.
- Accidents, incidents and emergencies.

Periodic routines may also be included in addition to the daily processes and activities. These may include preventive maintenance, plant upgrades, shutdowns or silent periods.

TABLE 2.2 Aspects and impacts

Item	Environmental Aspects (activity or process)	Environmental Impact	Caused By	Effect

TABLE 2.3 Local and global impacts

Local Impacts	Global Impacts
Surface water pollution	Global warming
Groundwater pollution	Ozone depletion
Local air pollution	Acid rain
Nuisance impacts	Resource depletion

Emergencies

Having prepared a list of the organization's activities and processes (aspects) and identified their environmental impacts (both positive and negative), it is useful to introduce a measure to determine their significance. Identify and prioritize those aspects that have a significant impact upon the environment. An environmental impact always occurs as a result of an environmental aspect. It is not always obvious that local environmental impacts often have a global impact, see Table 2.3.

Aspect and impact classification rating

For ease of explanation a five-point rating scale is used to classify each aspect's environmental impact as major, high, moderate, limited or minimal. The measurement criteria for each impact classification would be as given in Tables 2.4 and 2.5. A simple formula for an aspect's impact rating would be:

Overall impact rating = Aspect classification rating × Impact classification rating

Calculating the levels of 'low', 'occasional' and 'moderate', etc., is an assessment based on existing environment agency pollution acceptance levels, the experience and organization size. In determining the level of resources used bear in mind the

TABLE 2.4 Aspect classifications

Rating	Aspect Classification	Aspect Criteria
1	Minimal	No emissions and no use of resources No hazardous material usage
2	Low	Low emissions and low usage of resources Occasional use of hazardous materials
3	Moderate	Moderate emissions and use of resources Moderate use of hazardous materials
4	High	High emissions and high use of resources High use of hazardous material
5	Major	Major emissions and major use of resources Major use of hazardous material

types of materials being used. The use of solder paste, particularly lead based solder paste used to secure components in Printed Circuit Board (PCB) assembly for example, would typically carry a high aspect weighting. Be aware also that emissions can be to land, air and water.

Aspect and impact analysis matrix

To complete the toolbox each activity or process is assigned some degree of probability to the likelihood of the aspects and impacts occurring. Classification of an aspect can be as per Table 2.6. The probability of occurrence can be measured from a score of 1 – a less than 20 per cent probability, to a score of 5 – an 80 per cent to 100 per cent probability of occurrence. The calculated probability of occurrence scores are transferred to the potential aspect/impact/probability matrix.

To determine whether the aspect has significant environmental impact multiply the probability and impact scores together. If the aspect score is less than nine the aspect can be considered not to be significant. If the score is 9 or greater the aspect

TABLE 2.5 Impact classifications

Rating	Impact Classification	Impact Criteria
1	Minimal	No noticeable environmental effect Effective control system already in place Well within discharge consent levels
2	Low	Low environmental effect Control system in place but could be more effective Well within discharge consent levels
3	Moderate	Moderate environmental effect Control system in place but could be more effective Control system in place but must be improved Occasionally just outside discharge consent levels
4	High	High environmental effect No control system in place Outside discharge consent levels

TABLE 2.6 Probability of occurrence

Score	Probability of Occurrence
5	80–100% probability
4	60–80% probability
3	40–60% probability
2	20–40% probability
1	0–20% probability

can be considered significant. It should be noted that each process or activity may have a number of associated aspects and impacts.

Understanding the analysis matrix

The aspects that have been identified in the survey are listed on the left-hand side of the matrix shown in Table 2.7. In the next column put a number 1–5 to represent the percentage of probability of the aspect occurring. The next set of columns ranging from 5 to -5 represent the scoring system for the aspect having a positive or negative impact upon the environment. The examples given in the matrix demonstrate firstly a negative impact, a chemical spillage. An organization using chemicals in its operating process is likely to view any spillage as serious; therefore a -5 score would be applied to this aspect of its operation.

In contrast the reuse of packaging materials received from incoming goods can be viewed as a positive environmental impact as the same packaging materials can be used for packaging outgoing products. This would realize a score of 5 to signal a positive environmental impact.

By multiplying the scores together the figure in the furthest right column determines the degree of impact each aspect has upon the environment. The larger the negative figure the larger the environmental impact. The larger the positive figure the less of an impact upon the environment.

An aspect can have both positive and negative impacts upon the environment. Using the packaging example, if a significant amount of paper packaging from incoming goods was not reused, the recycling of this packaging waste would be a positive aspect but the money spent to have the packaging recycled may be a negative aspect.

Departmental analysis matrix

Without organizing larger businesses into separate departments it is virtually impossible to develop meaningful business definitions or meaningful and effective environmental management plans. It is important to stress that separate parts of

TABLE 2.7 Potential aspect/impact/probability matrix

Aspect	Probability of Occurrence	Good Score		Impact							Poor Score		Total
		5	4	3	2	1	0	−1	−2	−3	−4	−5	
Chemical Spillage	5											×	−25
Exhaust Fumes	5								×				−10
Product Packaging	5		×										20

the environmental management plan must be developed for every department in the organization.

The departmental analysis enables each part of the business to have its own business definition and an example is shown in Table 2.8. For every department a business definition should specify the following elements:

- The environmental objectives for each department.
- The resources available to meet the stated objectives.
- The technology to be utilized in product or process changes.

The use of a chart helps to visualize the implications and magnitude of the opportunities and threats. Specific values at the foot of the chart correspond with the total scores for each factor on the potential impact/probability matrix and indicate the potential implications of each factor on the organization or department.

Having created the register of aspects and impacts, attention can now be turned to the creation of the environmental legislation register.

Register of legislation

As a result of the external audit, a list of current legislation and regulations which has an immediate, or potential, impact on all business activities can be compiled.

Having assessed the organization's operational activities and the environmental implications, it is now possible to formulate an EMS. This process begins with the creation of a list of objectives and targets, and the development of a set of management and operational procedures. It is essential to be fully familiar, and compliant, with the regulations that apply to the organization. A fundamental part of the planning and implementation process is the creation and ongoing maintenance of a register of legislation and regulations.

TABLE 2.8 Departmental aspects and impacts analysis matrix

Departmental aspects and impacts analysis

Give rating from 1 (lowest or poor) to 10 (highest or excellent)

Key environmental factors	Weighting factor	The business		Department A		Department B	
		Rating	Adjusted rating	Rating	Adjusted rating	Rating	Adjusted rating
Waste generation	50	8	400	3	150	7	350
Energy usage							
Air emissions							
Total							

The process and activities review process enables an organization not only to demonstrate that it is aware of its existing legal and regulatory obligations but also can point to a mechanism for ensuring continued compliance with all new or pending legislation. A register of legislation is required to be maintained of those regulations that apply to the environmental aspects of an organization's activities. The register lists the regulations under the following headings:

- Item: A specified serial number is used for recording and sorting purposes.
- Regulation: Title and brief details of the regulation.
- Issued: Details of regulation issuing authority.
- Dated: The regulation issue date.
- Applicability: Details of the operations, activities or processes that are subject to the regulation.

The EMS guidelines require that a register is kept of the various items of legislation and regulation that are applicable to the organization. The scope of the register must include:

- International and national statutes and regulations.
- Requirements of specified permits, government documents and agreements.
- Contracts and other documents that may create binding legal obligations of relevance to environmental management.
- Requirements to each of the environmental aspects.

The register must be readily accessible by all staff members who need to be aware of all legislation and regulation that affects their activity. The register must be continually updated and reviewed and sources accessed to ensure continual awareness; typical source publications include:

- *Croner's Environmental Management.*
- *Environmental Business Journal.*
- *Journal of Environmental Law.*
- *The ENDS Report.*

Objectives and targets

The next stage of the process is to identify the objectives and targets for the EMS. When setting and reviewing these objectives and targets it is worth considering a number of influences within the organization, and these can include:

- Legislation and compliance requirements.
- Significant aspects – arising from the aspects and impacts analysis.
- Technological options – using the most appropriate technology.

- Financial – operational and energy efficiencies.
- Stakeholder interest.

For clarification purposes, an objective is defined by the ISO 14001 standards as:

> An environmental goal, arising from the environmental policy that an organization sets itself to achieve and which is quantified where practicable.

The achievement of, or the attempt to achieve, environmental objectives is the way an organization can improve its environmental performance. The objectives set do not need to be quantifiable but they do need to be realistic, identifiable and achievable. A target is defined by the ISO 14001 standard as a '... detailed performance requirement, quantified where practicable, applicable to the organization, or parts thereof, that arises from the environmental objectives and that needs to be set and met in order to achieve those objectives.' The first set of objectives and targets should reflect the introduction of the new system. As these short-term objectives are achieved so new objectives are set, thereby creating a dynamic set that takes the organization through the process of continual improvement. Therefore, a conservative approach is wise when estimating savings for the first set of targets. Until the EMS has been operating for a full year it is not easy to determine exactly how much energy has been saved or how much material has been reused. Examples of objectives and targets are given in Table 2.9.

The objectives of the EMS should encourage all employees to operate as part of an environmentally aware organization and to demonstrate this as part of an effective EMS.

Types of objectives

There are essentially three types of environmental objectives that can be established by an organization: compliance objectives, measurement objectives and quantified objectives.

Compliance objectives

These types of objectives are generally used in the early stages of EMS implementation to gain effective management control over those activities within an organization that are subject to legislative requirements, i.e. waste management, water discharge.

Measurement objectives

Until some initial baseline levels have been calculated it is not possible to establish, meaningful, measurement objectives. And, until a period of time has elapsed (one year) to aid assessment of activities it is also difficult to set reasonable and achievable levels against which to assess improvement.

TABLE 2.9 Examples of objectives and targets format

Item	Objective	Target	Method	Status
01	Completion of draft EMS documentation.	Completion of draft EMS documentation, including manual, procedures, logs and forms.	Follow requirements of ISO 14001 and any available examples.	Complete
02	Completion of final standard EMS documentation.	Issue of final standard EMS documentation.	Documentation to be finalized, reviewed, approved and issued following review of drafts.	Ongoing
03	Accreditation assessment preparation.	Successful completion of stage assessment.	Follow requirements of ISO 14001.	Ongoing
04	Premises clean-up. To include vehicles, tyres, containers, scrap items and oil contamination of soil and bund.	Removal of all potential contaminants and inherited contamination.	Liaison with local authority, clean-up agencies scrap dealers and vehicle owners.	Ongoing
05	Tracking and recording waste disposal. Recycling paper and cardboard.	Recycling and reusing as much paper and cardboard as possible.	Identification of suitable haulier and use of Waste Collection Log.	Ongoing
06	Reduction in office and workshop paper and cardboard usage.	Noticeable reduction in waste paper and cardboard output.	Publicity and placement of marked boxes throughout company. Personnel encouraged to reuse where possible and/or to avoid purchase.	Ongoing
07	Reduction in office and workshop plastic waste output.	Noticeable reduction in waste plastic output.	Publicity and placement of marked boxes throughout company. Personnel encouraged to reuse where possible.	Ongoing
08	Reduction in workshop materials usage and waste output.	Reuse of material and minimizing of waste output.	Publicity and placement of marked boxes in company workshop. Personnel encouraged to reuse where possible.	Ongoing
09	Reduction in energy consumption.	Reduce energy consumption annually by 10%. To include electricity and heating fuel.	Personnel encouraged to make savings where possible. Improve building insulation.	Ongoing
10	Reduction of hazardous material and substances.	Noticeable reduction in use of hazardous material and substances.	Personnel encouraged to identify and utilize environmentally friendly alternatives.	Ongoing
11	Reduce possibility of the emission of contaminated air.	A cleaner atmosphere.	Replacement of inefficient production or processing plant.	Ongoing

Quantified objectives

Similar to a measurement objective, a quantified objective cannot be calculated until baseline information has been collated and evaluated. Quantifiable objectives specify the performance requirements that need to be realized in order to assess whether the objective has been achieved.

Corporate environmental plan targets list

The primary objective is to operate and maintain the organization in a manner consistent with the best environmental practices, taking account of responsibilities to customers, staff, suppliers and the community at large. For this reason these objectives and targets are directly incorporated into the corporate environmental plan derived from this work. The plan should therefore be open-ended so those new or revised objectives can be added to, or changed within the programme. In the early stages this would be expected to be on a fairly regular basis.

The list of targets that is incorporated into the corporate environmental plan has been designed to be as simple as possible while showing complete details of goals, methods and responsibilities. The details of the active corporate environmental plan should be contained in an environmental documentation folder (EDF) for ease of reference.

The corporate environmental plan target details are contained in the targets list under the following headings:

- Item: A unique item number is assigned to each objective; this acts as a reference in reports or reviews.
- Objective: The specific objective is described and detailed. Some objectives may require more information; if so, this is provided separately and a copy placed in the relevant objective's folder.
- Target: (Example) Reduce energy consumption by 10 per cent.
- Goal: In this context, a goal is the projected target date for completion of the objective. There needs to be flexibility built into these dates; consequently, the dates may be altered following a progress review, as may the objectives and targets.
- Achieved: This is the actual objective achievement date.
- Responsibilities: Each objective has a person or persons designated as being responsible.
- Comments: This area is reserved for any pertinent comments.

Organization and responsibility

Having identified what the organization wants to achieve in terms of targets and objectives, the level of employee involvement should be specified, including an individual's specific tasks and responsibilities. Verification should also be made of the type and level of resources available to assess how achievable the objectives and targets are. Details of the training and communication programmes designed to ensure staff awareness and education should also be recorded.

Environmental operating procedures (EOPs) and environmental management procedures (EMPs)

Operational and managerial procedures are two types of environmental management procedures (EMPs). The establishment of environmental procedures is the main way in which environmental objectives and targets can be delivered. They are the basis on which the organization can be sure that the requirements of the ISO 14001 standard are being adhered to as they specify the way that any and every particular activity should be undertaken. All of the procedures in the corporate environmental plan are written to gauge the effective operation of the management system. Of the two sets of procedures, the operational procedures put processes into action that, eventually, allow the objectives and targets to be realized. The management procedures provide the controls. They ensure that the EMS is monitored continually and, where necessary, corrected and improved over time. Should a part of the EMS fail, the management procedures give guidance for reporting and correcting the failure.

Examples of EMPs are given in Boxes 2.1, 2.2, 2.3 and 2.4 and conform to ISO 14001 requirements and offer possible bases for procedure formation; this is discussed in further detail in this section. The five major EMPs are as follows:

- Control of Nonconformity.
- Management Review.
- Corrective Action.
- Document and Data Control, and
- Internal Audit.

Box 2.1   Example of management procedures

Management procedure: Control of nonconformities

Purpose

This procedure describes how nonconformities are controlled and reported. The procedure is outlined in Figure 2.3.

Scope

This procedure establishes the way in which nonconformities are classified, recorded and evaluated.

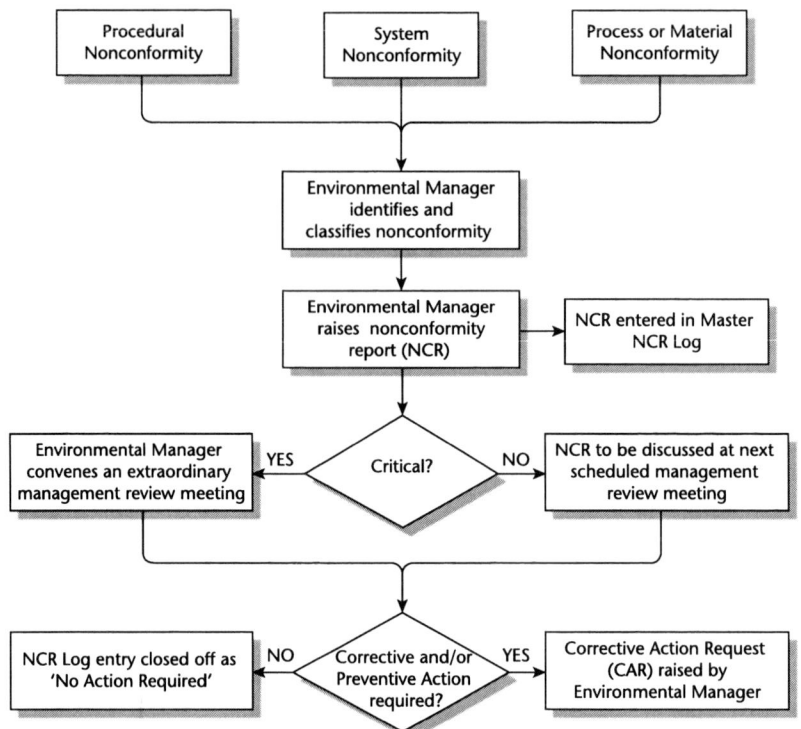

FIGURE 2.3 Control of nonconformities

Responsibility

The environmental manager is responsible for documenting nonconformities, presenting them at management review and ensuring that follow-up actions are undertaken.

Procedure

There are three types of nonconformity:

1. Procedural nonconformity – where personnel have failed to follow EOP or other forms of instruction contained within the EMS (typically identified during internal audit or periodic review by the certifying agency), or if the nonconformity is an accident or a hazardous occurrence. The most obvious example is where someone has failed to complete a new product introduction (NPI) form when introducing a new product into the company.
2. System nonconformity – where a fault or inadequacy is identified in the EMS and change is required to assure the safety and protection

of the internal and external environment. Using the NPI example again, the procedure for introducing new products into the company may not have been created.

3. Process or material nonconformity – an accident or occurrence not related to either of the above types of nonconformity (typically involving an unforeseen breakdown or failure of equipment, process or facilities). An example of this type of nonconformity may occur when a piece of machinery fails and an oil or diesel spillage creates a potentially hazardous situation from the threat of fire or water and/or land contamination.

Nonconformity identification

Nonconformities may be identified by any company employee, or by anyone such as a subcontractor who may be working for the company.

Nonconformity reporting

The nonconformity reporting chain is achieved through the use of existing management structures and the established interface.

It may not be feasible or practical for initial reporting of a nonconformity to be done in writing. In many cases a verbal report will be made to the environmental manager who will carry out an initial investigation and then produce a written report. Unless it is impossible, all reports should be put in writing eventually, regardless of the initial method of transmission.

All final nonconformity reports submitted to the environmental manager should conform to the reporting form layout distributed by the environmental manager and include details of immediate actions taken, or to be taken.

A nonconformity can be either critical or non-critical. A critical nonconformity has a direct and immediate effect on safety or protection of the environment. For example, a faulty piece of equipment might be identified or an EOP or other form of instruction might be found to contain an error that could affect the environment.

Any other type of nonconformity is, by definition, non-critical and the environmental manager has the task of deciding whether a nonconformity is critical or non-critical.

Environmental manager actions

When details of a nonconformity are received, the environmental manager shall produce a nonconformity report (NCR) in which:

- the type of nonconformity is identified;
- it is classified as critical or non-critical; and

- the nonconformity is described in sufficient detail to allow management to identify the appropriate corrective action.

Every NCR should have an issue date and be identified, for example, with a four-digit number (e.g. 0001). The environmental manager should retain copies of all NCRs and maintain a master log of them. The log identifies NCRs by issue number and date, and shows the date of the management review meeting at which the close-out actions and responsibilities would have been discussed. Each NCR is closed-out in the log by the allocation of a corrective action request (CAR) number or numbers or when management review decides that no action is to be taken.

If the nonconformity is designated as critical, the environmental manager will call an extraordinary management review meeting as soon as possible. Temporary remedial action may also be taken immediately and detailed on the NCR. If the nonconformity is non-critical, the NCR will be presented at the next scheduled management review meeting.

Box 2.2   Management procedure – Management review

Purpose

This procedure describes how management shall review the EMS at regular intervals to ensure it continues to satisfy the environmental policy.

Scope

This procedure establishes the conduct by which a management review is undertaken.

Responsibility

The managing director is responsible for ensuring that the EMS is reviewed according to this procedure.

Procedure

The review should be conducted by means of a pre-planned meeting chaired by the managing director, and consisting of the environmental manager and one (more if required) member of the EWG. Management review meetings should take place at least once in each six-month period. These meetings can be held annually depending on the company circumstances, but six monthly is recommended The environmental manager shall advise the managing director when to announce the precise date of the meeting.

Unscheduled, extraordinary review meetings may also be held should circumstances dictate, particularly in the event of fire and/or floods etc.

The environmental manager is responsible for:

- Convening a management review meeting and drafting an agenda for approval by the managing director.
- Maintaining the environmental records that provide input to the management review.
- Ensuring that minutes of the review and all agreed actions are recorded for subsequent approval by the managing director.
- Preparing and updating an EMS audit schedule for approval by the managing director.

The environmental manager should maintain an EMS audit schedule showing:

- The frequency and timing of audits.
- Specific areas and activities to be audited.
- Identity and qualification of auditor(s).
- Auditing and reporting criteria.

The EMS audit schedule shall be regularly drafted and/or updated by the environmental manager for approval by the managing director. The redrafting should be carried out at least once in any twelve-month period. Every company employee is responsible for compliance with the EMS audit schedule by participation, assigning resources, etc., as required.

The management review agenda shall typically include:

- Review of the minutes of the previous meeting and actions carried forward requiring close-out.
- Results of EMS audit activity.
- Review of the environmental manager's summary report of nonconformities since the previous meeting.
- Review of organizational management procedures.
- Review of administrative procedures.
- Review of EOPs.
- Review of personnel responsibilities and authority.
- Review of documentation and record keeping.
- Review of and adherence to EMS policies, procedures and instructions.
- Need for additional familiarization or on-the-job training.

- Results arising of analysis of any critical nonconformity such as personal injury, equipment damage or pollution incident.
- CAR close-out reports produced on non-critical and critical nonconformities since the previous meeting.
- Corrective action taken on operational defects or procedural amendment in the EMS and further measures to improve its effectiveness.
- Recommendations from employees and EWG meetings for measures to improve EMS effectiveness.
- The degree to which the environmental policy continues to meet the company's current objectives and statutory compliance requirements.

The environmental manager shall revise the EMS audit schedule according to the actions arising from the management review and present it to the managing director for approval. The results of the review should be brought to the attention of those persons responsible for implementing the changes proposed.

All actions arising from a management review should be closed-out and signed off by the managing director.

Continuous improvement

Monitoring and measuring the existing environmental management plan is a necessary requirement for ISO 14001. A similar requirement for the standard is a programme or system that facilitates continual environmental improvement. It is too simplistic to set a number of targets and objectives, achieve them, and sit back and smoke the cigar. Once existing targets and objectives have been achieved, more need to be set. These future targets and objectives may take the form of new targets based on existing objectives, for example a further 5 per cent saving on energy usage, or completely new objectives and targets.

As mentioned at the outset, energy savings or waste minimization levels can be too ambitious; set achievable annual goals. It is better to have achieved the extra 5 per cent savings targets each year than a 25 per cent one-year target. The latter may well be difficult to achieve in one year, and yet failure to do so may have a negative effect on morale and the programme of continuous improvement.

Box 2.3   Management procedure – Corrective action

Purpose

This procedure describes how corrective and preventive action is initiated and maintained within the EMS, and is outlined in Figure 2.4.

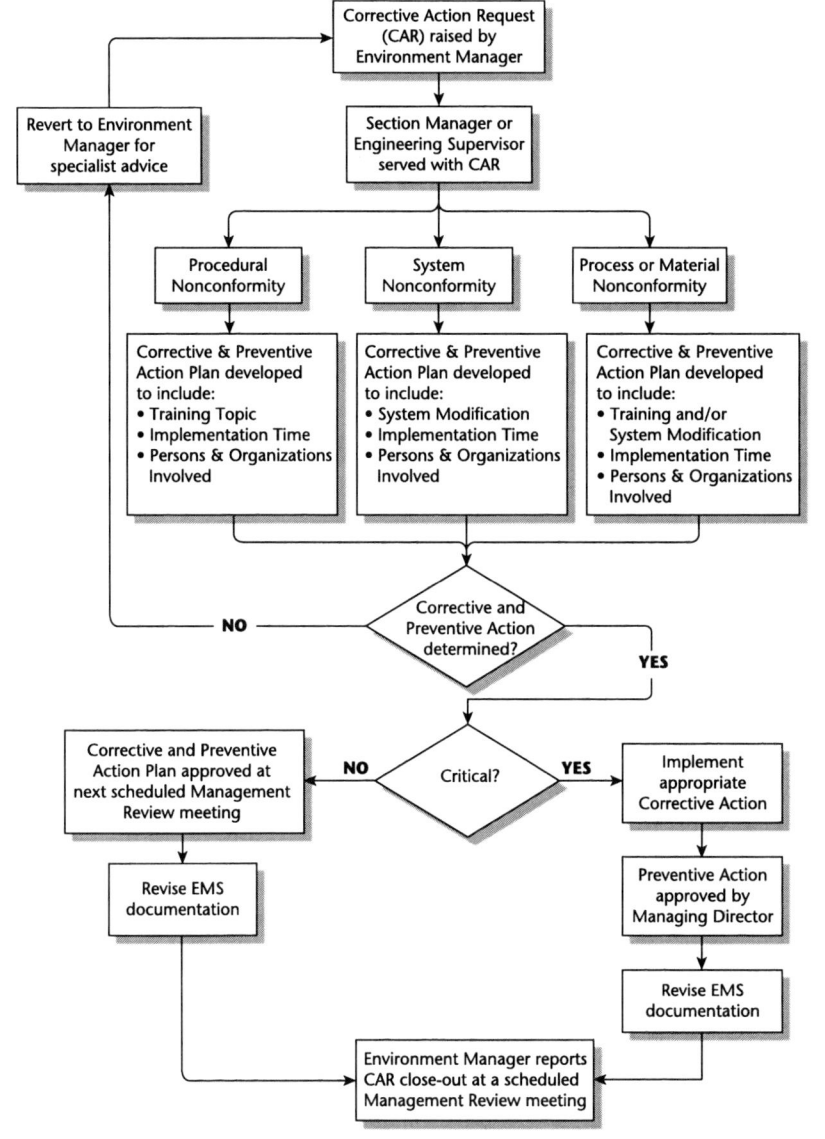

FIGURE 2.4   Corrective action process

Scope

This procedure establishes the way in which corrective and preventive action is documented and reviewed.

Responsibility

The environmental manager is responsible for producing a CAR for any corrective action decided at a management review.

Procedure

The CAR will provide brief details of the nonconformity to which the corrective action relates and will provide information on what needs to be done. It identifies the addressee – the person required to complete the corrective action – and includes a timescale for completion of the corrective action.

The environmental manager retains copies of all CARs issued and maintains a log of them which shows:

- CAR number.
- NCR relating to the CAR.
- Addressee.
- Date CAR was raised.
- CAR status (actioned or to be actioned).
- CAR close-out date.

Close-out of corrective actions

CARs relating to a procedural nonconformity are closed-out when the auditor who raised the NCR has reviewed the corrective action and accepted that it addresses the requirements of the NCR. If the original auditor is not available, the environmental manager may close-out a procedural nonconformity.

CARs relating to cases of system or material/process nonconformity are closed-out when the environmental manager has reviewed the corrective action and accepted that it satisfies the requirements of the NCR.

The environmental manager will produce a CAR close-out report on all closed-out CARs. This report will briefly summarize the NCR, the corrective action required and the actual action taken. The environmental manager will retain copies of all CAR close-out reports.

Copies of CAR close-out reports will be distributed to management at the next scheduled management review meeting.

Box 2.4 Management procedure – Document and data control

Purpose

This procedure shown in Figure 2.5 describes the control of documents and means of electronic data control (including indexing, data storage and protection, and archiving) that are essential to the effective operation of the EMS.

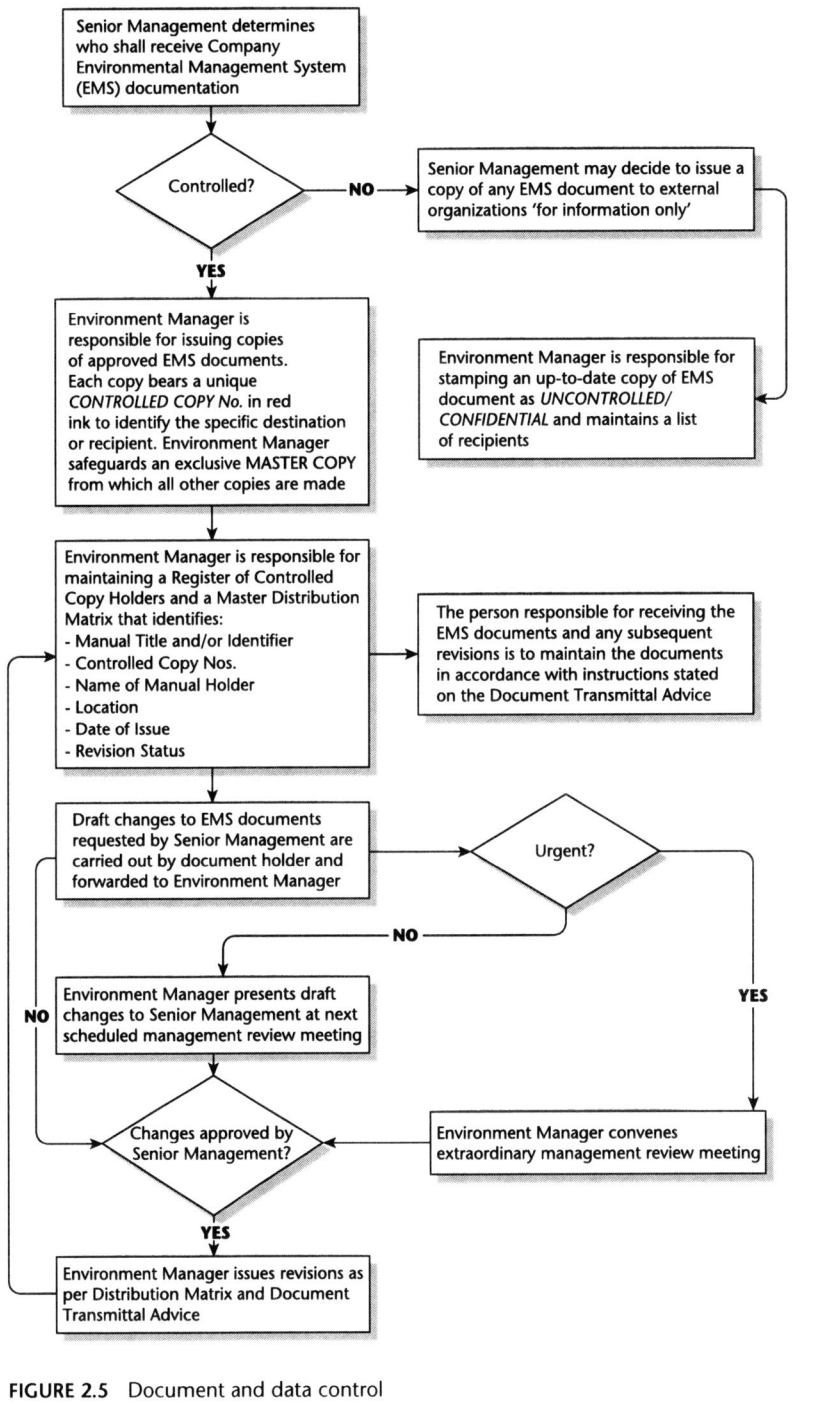

FIGURE 2.5 Document and data control

Scope

This procedure should apply to the following:

- Environmental management procedures (EMP).
- Environmental operating procedures (EOP).
- Forms and reports.
- Work or other instructions.
- Any other environmental documentation.

Responsibility

The issue, amendment, approval and distribution of all EMS documentation are to be carried out under the direction and authority of the environmental manager. Document holders are responsible for the maintenance of controlled documentation and for ensuring that all obsolete documentation is returned as instructed.

Procedure

All documents relevant to the EMS shall be reviewed, maintained and controlled and ensure that:

- Documents and their subsequent revisions are reviewed and approved by the environmental manager prior to use.
- The issue and revision status of the documented EMS is known to all users through the use of controlled documents.
- Persons issued with controlled documents are aware of their responsibility for maintaining the revision records.
- Up-to-date issues of appropriate documents are provided at any locations where critical operations are performed.
- All copies of obsolete and superseded documents are withdrawn from use to be destroyed or archived.
- Any obsolete documents retained as records are identified suitably and that a master list is maintained of all documents and data under control.

Control of company originated EMS documentation

All such documents shall bear a control header design (as used in this procedure) which shall include (in designated form and location):

- A corporate identifier (as appropriate), document title, chapter number and title.
- Issue status such as draft or the date (expressed as month and year).
- Revision number, starting at 00 and increasing sequentially in units.

- Page number, or chapter number and page number (expressed as x or x–xx) or page number and total number of pages per chapter (expressed as x/xx).
- Forms shall have a unique numbering sequence that indicates their function.

Distribution matrix

The environmental manager shall maintain a distribution matrix that records the following information about all documentation relevant to the EMS and any changes affecting the location and/or status of individual publications or copies shall be recorded on the distribution matrix:

- Document title.
- Controlled copy number.
- Location/holder.
- Revision status.
- Issue date.

Document transmittal advice

When a publication or document is issued, or if a revision is issued by the environmental manager, it shall be accompanied by a document transmittal advice form. This form is addressed to the person who will be receiving the new publication or who is responsible for revising an existing publication.

If the form accompanies a new publication, the recipient is required to signify receipt and respond in accordance with the instruction given on the form. The revision record is to be completed and signed in accordance with the document transmittal advice form instructions that also detail the response required.

Revision record

Other than forms and minor documents, company originated EMS publications are prefaced by a revision record. This is a record of revisions to the specific publication and, therefore, provides essential evidence that a publication is being maintained up-to-date. It is the responsibility of the person having charge of the publication to ensure that the record is maintained correctly and accurately.

Control of electronic data

All electronic data that originates from within the company must be identified, indexed and stored in accordance with current company instructions. Access to company computer systems should be granted only to those persons authorized. Access to computer files should be similarly controlled. Such precautions can minimize the possibility of accidental or deliberate corruption of essential information, or electronic infestation that might corrupt or destroy data.

Control of documents from external sources

The environmental manager should be required to maintain an up-to-date list of all externally supplied publications and should also be responsible for the purchase and supply of such publications.

Each publication should be given a unique number or code upon receipt. The date of issue should be clearly stated. Obsolete copies should be withdrawn from service and marked 'SUPERSEDED' prominently on the front cover.

As specified in other procedural guidelines, the start of any procedural prescription should specify the purpose of the procedure, its scope, the person responsible for its execution and its administrative process.

Reviewing EMS effectiveness

An important part of establishing and maintaining the effectiveness of the corporate environmental plan is ensuring that the organization's significant environmental impacts are being controlled on a daily basis. The process of monitoring and measuring ensures information is received to ensure that this control and continuous improvement occurs. Monitoring and measuring should occur at regular intervals within the stages of implementing the EMS. Once the objectives and targets are set, responsibility has been allocated and the operational and managerial procedures are in place, the opportunity arises to monitor and measure the EMS in preparation for the EMS audit, which evaluates the effectiveness of the EMS.

The following sections describe some of the documentation and review processes used to monitor and measure the progression of the EMS. This includes:

* Written annual environmental management reports.
* Environmental plan reviews and control processes.
* Identification of environmental performance indicators.
* Opportunity and threat analysis.
* The evaluation of previous corporate environmental plans.

Environmental management reports

Presenting an annual environmental management report provides an opportunity to review how well targets and objectives are being achieved, and assesses the way in which the environmental management planning function is intended to operate. The environmental report has been identified as a key tool in communicating environmental performance to employees and stakeholders and therefore improving staff morale and corporate image (Ramus, 2002). The report functions as a monitoring and measuring device, by providing a point of reference, or benchmark, against which to measure the progress made since setting the targets and objectives. An outline of the control and evaluation mechanisms helps to assess or evaluate current and previous environmental management plans. The issue and control of documentation is important to the effective monitoring and control of environmental procedures. Quality control documents may already exist within the current control procedures of the organization and these should be used to avoid repetition and extra work.

Environmental plan review and control aims

Four of the most common aims of the plan review and control processes are:

- To permit problems or developments that do not match planned or budgeted schedules to be identified early and addressed if required.
- To identify their causes and act to nullify their effects.
- To provide input into the ongoing environmental management function of identifying environmental management aspects and impacts.
- To act as a performance indicator and stimulus for environmental management personnel.

Control mechanisms may include or involve the development of:

- Performance criteria and standards.
- Acceptable ranges within which these criteria and standards are deemed to have been satisfied.
- Procedures which provide suitable and reliable measures of results.
- The means to compare the results achieved with the standards and criteria set.
- Systems which enable effective corrective measures to be taken.
- A reliable means of forecasting outcomes.

The degree to which budgeted targets for each product within the corporate portfolio have been realized, measured by value and volume, should ideally be checked monthly. A system – an existing budgetary system would suffice – must be established to enable the production of reliable and useful data as a matter of routine and as and when it is needed.

Data must be available at the appropriate level; aggregated figures showing the savings derived through implementation of the EMS, for example, detailed by department if required. Budget responsibilities need to be delineated clearly – for instance, a clear budget responsibility should be specified for the environmental management function which encourages accountability.

Plan review meetings can also function as occasions on which certain kinds of information can be disseminated. For example, environmental management research data which may have been commissioned on behalf of one department may not easily 'trickle across' in less formal situations.

Controls also operate over new product development, with specific reviews of the development of such products or services. Continued shortfalls against projected performance should trigger revision or remedial activity.

Environmental performance indicators

As the aspects and impacts of an organization vary according to its business activities, so the creation of environmental performance indicators (EPIs) varies. The following sections list possible EPIs that should be measured:

Resource usage measure

This measure provides insight into the consumption of energy, water and other resources at the organizational, departmental and process levels. Examples would include:

- Tonnage of raw materials used per unit of production.
- Percentage of recycled or reused materials per unit of production.
- Energy consumption per unit of production.

Emissions/waste measure

There are no generally accepted guidelines for measuring or reporting emission levels. However, once set, under the ISO14001 standard guidelines, organizations are required to demonstrate continuous improvement against this benchmark. Monthly spreadsheets and graphs provide suitable evidence to demonstrate the following:

- Emissions of effluent per unit of production.
- Percentage of paper or cardboard recycled or reused per unit of production.

Environmental impact measure

This measure assesses the impact the organization's activities have upon the environment. For example transport organizations may measure the emissions of CO_2 from vehicles (emission/waste measure) but may not measure the degree of impact CO_2 has on ozone depletion.

Environmental risk measure

The increasing amount of environmental legislation and regulation is a key driver for most organizations seeking to reduce the probabilities of an environmental accident such as a tanker oil spill. It is too simplistic for an organization to arbitrarily state that there is a 15 per cent chance of a major incident in the next three years without undertaking an environmental risk assessment exercise.

Management systems measure

The degree of management input into a new environmental system can be a useful measure of innovation. The introduction of new manufacturing procedures may create significant cost savings from improved processes or create new products or service opportunities.

Customer measure

Customer measurement should pervade all organizations. The organization must:

* Identify its target customers.
* Convey the needs of these customers.
* Show how its products and services satisfy these needs.

Competitor measure

It is also useful to conduct a competitor analysis, which helps understand relative strengths and weaknesses. This involves the identification and weighting of operating factors to determine their relative importance in the industry. When all weights are added together the total should add up to 100. Each success factor must then be assessed against other major competitors on a scale from one (a low or poor competitiveness rating) to ten (a high or excellent competitive rating). Adjusted competitiveness ratings are calculated by multiplying the weighting factor by the individual organization's ratings. Total organization competitiveness ratings are calculated by adding together the adjusted ratings for each success factor for each organization. To determine competitor ratings, use the departmental analysis table (Table 2.8) and change the departmental classification to competitors.

Efficiency measure

There can be many levels of corporate activity where efficiency measures could be applied. Taking a holistic view of the organization the ratio of total inputs – such as energy and resources to total outputs – such as products and services. At the operational level the measurement of energy efficiency is also important. It is a simple process to calculate the bottom line energy cost with the following formula:

$$\frac{\textit{Total cost of energy bill}}{\textit{Number of kWh used}} = \textit{cost per kWh}$$

Another method of measuring efficiency is the plant load factor (PLF) and is represented by the following formula:

$$\frac{Annual\ consumption\ of\ kWh \times 100}{Maximum\ demand\ (kW) \times Hours\ in\ the\ year}$$

Improving the PLF reduces the supply price and therefore reduces manufacturing costs.

Financial measure

This measure can be as simple as the recording of costs associated with dealing with an environmental issue or implementing an EMS. This may be introduced at the time of setting environmental budgets and together with a method of exception reporting as when there are significant over or under spends.

Impacts measure

Direct measures of emissions or waste indicate the impact of operational activities upon the environment. For example, samples of discharged water can be collected and independently analysed. The levels of the emission found provide benchmark levels of contaminants that can either be reduced or eliminated completely.

Opportunity and threat analysis

A good way of thinking about the implications and magnitude of impacts is to reflect upon them as opportunities or threats by using Figure 2.6. Specific values on the bottom of the chart correspond with the total scores for each factor on the

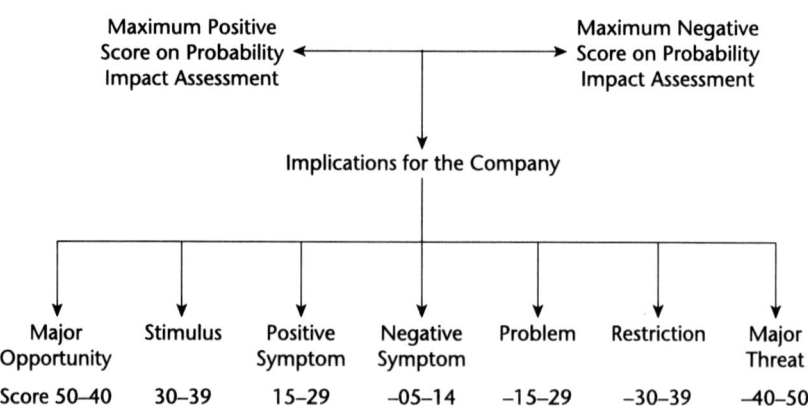

FIGURE 2.6 Impact opportunity or threat analysis

potential impact/probability matrix and indicate the potential implications of each factor on the organization or department.

The evaluation process of previous management plans

In order to regulate and control the environmental management plan effectively, use environmental reports or other organization information to locate the causes of under-performance precisely, and to spot those things which are working and which the environmental management plan is therefore getting right.

Two main types of measure can be employed: 'Environmental management costs to sales ratios', and 'Customer tracking'.

Environmental management costs to sales ratios

This ratio relates the amount spent on environmental management activities to the sales that have been achieved by the organization. These are important, but broad, measures which must be interpreted carefully, and provide a detailed check on environmental management expenditure. For example consider that acceptable annual environmental management costs should equate to 1 per cent of sales revenues.

Customer tracking

As with marketing and advertising initiatives monitor how customers feel about the organization's products and its environmental management activities. Such tracking includes a wide range of qualitative and quantitative measures of customer reactions taken from panel data, internal records of customer complaints, sales force reports, focus group interviews and surveys.

Because each business is different, it is difficult to outline a standard system for evaluating performance. The following list, however, may help:

* What has happened since the previous environmental management plan?
* How this compares with the timescale or programme indicated in the previous environmental management plan?
* How it measures up to the intended progress?
* Why this has happened?
* What extra costs (if any) have been incurred?
* How it fits into the budgeted figures?
* What actions are required as a result?

Of course this may be the first environmental management plan and there may be little or nothing to compare it to. If a previous environmental plan exists, a simple review would assess whether previously specified targets and objectives had been achieved. If they have, set some new ones, if they have not, find out why not. The ethos of continuous improvement on which the ISO 14001 standard is based must be evident within environmental plans. That said, do not make them a huge burden for the organization to achieve.

The production of corporate environmental annual reports means little if the organization does not have the capability to measure its performance. Having prepared a list of activities and processes (aspects) and identified their environmental impacts (both positive and negative), it is useful to introduce a measure to determine their significance and to set benchmarks to determine whether progress is being made.

The implementation of an EMS is of no benefit to an organization unless it can demonstrate to external, independent observers and stakeholders that improvements are being made. The collection of detailed data from clear EPIs, over time, provides a vehicle with which an organization can set benchmarks and demonstrate through catalogued evidence that continuous environmental improvement is being made.

Environmental management system (EMS) audit

The environmental audit that occurs at the early stages of implementing an EMS is concerned with assessing the external and internal environmental issues of the organization. This audit assesses whether the performance of the EMS conforms to the planned objectives and the requirements of the ISO 14001 standard. Any recommendations for change to the system generally follow the system audit. Findings from the audit are discussed and action required noted at the management review.

The next section explores the audit procedure for testing the environmental management system and to establish a management review format to determine further development or changes to the system. Figure 2.7 demonstrates the number of elements involved in controlling the efficient functioning of the EMS.

The EMS audit helps determine whether:

- The business activities are conforming to the requirements of the EMS.
- There is employee awareness, together with procedural familiarity and compliance.
- There is operational relevance, accuracy and effectiveness of the EOPs and EMPs.
- There is proper determination of the EMS adequacy by senior management.

Audit plan and procedure

The way in which internal audits are scheduled, planned, carried out and recorded is detailed below. It is the responsibility of the environmental manager (or other dedicated individual), together with that of the working group for the development of an audit plan and schedule. A suitably qualified auditor should be used to undertake the audit.

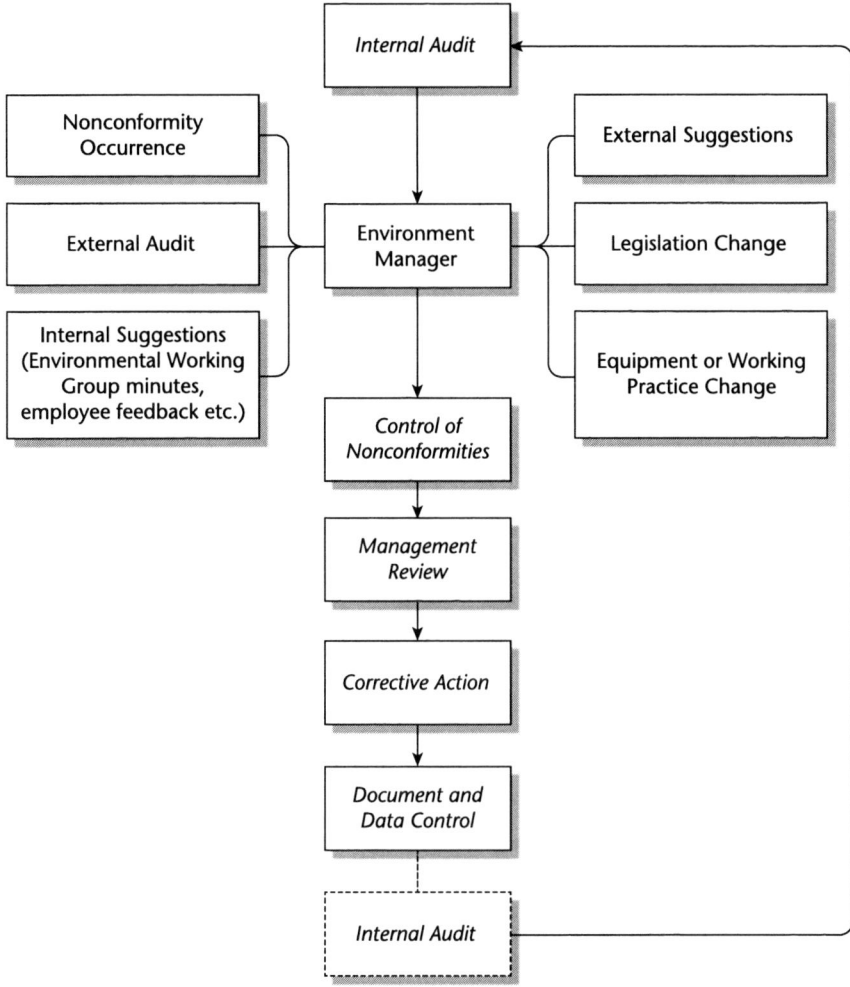

FIGURE 2.7 Environmental management procedure overview

This final audit requires an audit of the complete system covering all operations of the organization. The best place to start is by devising a schedule of all the audits that have to be undertaken. The audit schedule detailed below shows that the audit functions to be undertaken are spread throughout the year at a frequency of either one or two per month. Some audits are more time consuming than others, so plan the schedule carefully. Schedule audits systematically so that they are all completed within one year. During the implementation phase of any EMS there can be economic trade-offs. One example of a trade-off may be the acceptance of low prioritization, or late response by some departments when implementing environmental procedures due to operational requirements. A late response may be considered preferable to no response.

Audit notification

Under the requirements of the ISO 14001 standard, regular audits of the EMS need to be carried out to ensure that it continues to meet its objectives and staff comply with management and operational procedures. Prior to any audit being undertaken, each person to be audited must be notified a minimum of one week in advance of the audit date. It may also be useful to post a list of all employees to be audited, including times and dates, on the organization notice boards. The notification of audits may seem like a minor piece of auditing etiquette, but the assessor requires proof of audit notifications and, if not identified, system nonconformity is the likely outcome.

The audit notification requirements can be very simple in format but must provide advance notification of the following:

- Who is to be audited?
- When and where does the audit place?
- Details of the topics to be covered by the audit.

In addition to the above requirements a brief note as to how each person is to answer the audit questions and where the answers can be located would also prove to be time well spent. It is always helpful to spell out the audit requirements clearly and in full. Audits can at times be viewed with suspicion and trepidation so time taken to reassure staff aids the process and improves results. Set a completion date for all persons to be audited. This ensures that all audits are completed quickly. Having a final completion date helps ensure that any time changes remain within the allotted time period.

Audit questions

Table 2.10 is an example of some audit questions that might be asked. The presentation is simple with space available for answers. Prior notification of any audit allows each person being audited the opportunity to dash to the nearest environmental manual and familiarize themselves with the relevant procedures.

When setting the audit questions choose whether to make them difficult or simple. The benefit of the simple option is that people feel less threatened with the task, and will be more inclined to take time to read the relevant procedures. Simplify the audit by pre-warning the individual of the systems and procedures to be audited. Due to operational pressures most employees appreciate more direction and less reading of procedures. This does not diminish the exercise. The audit procedure will most likely be a new system and therefore a new learning experience for everyone. Communication has a strong role to play here to encourage everyone to understand what is required of them and how the new system works.

TABLE 2.10 Examples of audit questions

Question
Who is responsible for the identification, handling and storage of all waste?
What procedure defines the methods that shall be used for the handling, storage and disposal of liquid and solid waste produced in the office, workshop and stores areas of the organization?
Who is responsible for documenting and managing the EMS?
Where would information regarding the way in which spillages of hazardous materials should be dealt with be found?
Who is responsible for the storage of all materials?
What is the name of the plan that is designed to be a dynamic and open-ended list of objectives and goals for the company to achieve in accordance with ISO 14001?
What is the correct method for the disposal of waste categorized as rubbish?
Who is responsible for the storage and maintenance of all records and completed forms that are part of the EMS?

During the audit process, the auditor examines:

- The organizational structure.
- Management and operational procedures.
- The workplace including layout and operations.
- How well the EMS meets the requirements of the organization's environmental objectives.

It is highly unlikely to successfully audit an EMS without incurring one or more nonconformities. In fact, the more nonconformities found, the better the system and the more satisfied the assessor. The audit process provides two key benefits: first, it demonstrates that the EMS works and, second and more importantly, it demonstrates the working of the management procedure to identify, report and action nonconformities.

Audit report

Having completed the audit, a supporting evidence base is required. The construction of an audit report is the next task. This provides the assessor with two pieces of evidence of a successfully working procedure and system. The first is the proof that an audit has been carried out in accordance with the auditing procedure. The second is proof that those employees audited have read the procedures and are familiar with the general requirements of their role in the EMS.

Box 2.5   Template for audit report

Auditee: Auditor:

Audit Date: Audit Location:

Audit Objective: Audit Report Number:

	Yes		No	
NCR				NCR Number:
CAR				CAR Number:
Follow-up Action Required? (see narrative)				Reference Number:

Narrative:

Signed (Auditor):   Date:

The audit report should consist of three elements:

- A cover sheet.
- The response to the audit questions.
- The conclusions of the auditor.

The audit is in effect the organization's annual medical check-up. The audit report contains details of individual environmental audits that have been conducted in accordance with the requirements of the EMS auditing procedure.

Management review

If the organization is to achieve continual environmental improvement, the environmental plan needs to be reviewed and examined regularly. The review should not be seen to be restrictive and, where possible, should encompass all aspects of the EMS. It should be viewed as a health check for the well-being of the system.

The senior management team undertakes an annual review of the EMS in order to ensure its continuing suitability and effectiveness. The reviews establish the need, if any, to change policy, procedures, controls, objectives or other relevant matters, taking account of the audit results, changing circumstances, including legislation and the need for continual improvement. The management review discusses every element of the EMS to ensure that each is working effectively. Future objectives and targets are identified and discussed during the review. The commitment of senior management is most in evidence at the management review. Every decision made by those senior managers present at the review is accepted by all and the managing director ratifies the minutes of the management review.

The existing policy statement also needs to be reviewed annually by the managing director who should update the contents in order to stress the commitment of the organization towards continually improving the environmental commitment of the organization.

A major outcome of the initial environmental review should be the formation of an environmental working group (EWG) headed by a senior manager or director. The EWG should be tasked with meeting at least once a month in order to discuss any matters that have arisen and to progress any ongoing environmental topics. Meetings can be arranged more frequently if the need arises, particularly if the plan is to introduce the EMS quickly.

Employee involvement in environmental matters is fundamental to the adoption and maintenance of a successful EMS. A number of ways in which employee involvement in the EMS could be improved should be proposed, including a survey for suggestions as to how the organization's operations could be improved in terms of environmental protection. In addition, all employees should be urged to highlight any process or operation that could have any adverse environmental impact.

Environmental management review minutes

Outlined in Box 2.6 is an example of the review meeting minutes. Minutes of meetings are necessary to provide evidence that the review has taken place and to ensure that issues raised are acted upon.

Box 2.6   Example of review meeting minutes

Date/Time:
Location:
Present:

Item	Topic	Action
01.	Initial meeting. Consequently, there were no previous minutes to review and no previous actions had been carried forward.	
02.	Reported on personnel audit progress and discussed the proposed audit schedule.	
03.	The two NCRs that had been submitted were discussed. CAR action had been initiated.	
04.	The use and effectiveness of EMPs was discussed. No problems or difficulties arose.	
05.	The use and effectiveness of EOPs was discussed. No problems or difficulties arose.	
06.	Personnel responsibilities were discussed and were considered adequate. No points arose.	

07.	Documentation and record keeping was discussed. Current standards were considered adequate, and no points arose.
08.	The need for additional familiarization and training was discussed. Current levels were considered adequate. No points arose.
09.	There were no critical nonconformities raised to date. No points arose.
10.	The two current CARs were discussed. Recommended actions would be initiated by the environmental manager.
11.	No employee or working group recommendations or suggestions have been received.
12.	The functionality of the environmental plan was discussed and it was agreed that no changes were required at this stage.

Minutes prepared by:

Authenticated: Environmental Manager

Minutes agreed: Managing Director

Extraordinary management review meeting

Any person within the organization can, and should, report an actual or potential environmental nonconformity. All nonconformities shall be reported directly to a senior manager, either in person or in writing. The senior manager discusses the report with the originator and assists in the accurate recording of the nonconformity. If a critical nonconformity is identified, (i.e. if it has an immediate and direct effect on safety or protection of the environment), immediate preventive action should be taken by a responsible person. Following consultation with the senior manager or director, an extraordinary management review meeting may be convened in order to approve further corrective action. Every system nonconformity outcome identified should be evaluated at a management review meeting where it is decided if corrective action is required or any prior action taken has achieved its aim. The management review process also examines the EMS as a whole to ensure that it continues to meet the objectives of the corporate environmental policy.

Corrective action

Corrective action is the action taken by a responsible person under the direction of a senior manager to prevent the recurrence of a system nonconformity or to improve the EMS. Corrective action is formalized by:

- Revision to the EMS documentation.
- Issuing the changed documentation.

- Initiating a follow-up audit to verify implementation and effectiveness of the corrective action.
- Internal audit.

The cycle of activities is completed when the corrective actions are closed-out under the internal audit procedure. The role of the senior manager in the cycle of nonconformity reporting, the implementation of corrective action, follow-up audit and document control is essential to the effective operation of the EMS.

Creating a corporate environmental management plan

The corporate environmental plan is the culmination of the auditing process. Having gone through the identifying, listing, analysing and prioritizing of the operational aspects and their environmental impacts, the emergent targets and objectives form the basis of the corporate environmental plan. Design the corporate environmental plan to be a dynamic and open-ended list of objectives and targets. Some of the initial targets and objectives set in the early stages of introducing an EMS are realized when accreditation of ISO 14001 has been achieved. One of the first objectives listed may be the achievement of an environmental management standard.

In short, some objectives and targets have completion deadlines, and when achieved these should be deleted from the corporate plan and replaced with new objectives and targets. As the management review is undertaken on an annual basis, operational activities change this and necessitate the introduction of new targets and objectives. The process of reviewing and restating of environmental objectives and targets offers the opportunity for continuous operational and environmental improvements to be made.

As with the previous stages of the process, this plan must be regularly reviewed and updated by the environmental manager in consultation with the EWG. The managing director undertakes a periodic review of the EMP. Periodic progress reports are circulated within the organization.

An environmental management plan should be easy to read and avoid jargon wherever possible. Most of the people reading the plan are not environmental management experts. If complicated terminology is used include definitions and explanatory notes. In some cases a glossary of terms is necessary and should be included as a separate section at the end of the environmental management plan.

Number each section and page so that the reader can easily navigate the plan and visual aids such as tables, charts, and diagrams are often the best way to convey complex information. Photographs of the products, processes and business activities can be used to give the environmental management plan life. Be candid and honest when conducting forecasts and writing the environmental management plan budget. Over-optimism can cause doubt in the reader's mind about the credibility and judgement of the author.

Once the first draft of the corporate environmental plan has been completed, cross-check it for any operational inconsistencies. For example, make sure the environmental management aspects and impacts and EMPs are in line with the organization's objectives and targets. At this stage check that the content of the environmental management plan is relevant to the reader and discard any surplus material. It is also vital that the objectives are communicated and wherever possible agreed with those who are charged with the responsibility of achieving them.

Typically, organizations have operational objectives affecting many areas of the organization, any one of which can affect the selection of environmental management objectives and strategies. The business objectives are likely to encompass more than simply finance and involve those of both a qualitative and quantitative nature. Some of the more common objectives for many organizations include:

Qualitative	*Quantitative*
Market standing/reputation	Profitability through materials recycling
Innovation	Greater production efficiency
Management performance	Profitability through energy efficiency
Public responsibility	Identification of new products
Organizational development	Efficient operational processes

When framing the objectives, detail how they are achieved. For example, an objective to increase market share with a low product price may be written as:

> Our objective is to decrease energy consumption by 10 per cent in the following year by conducting an extensive energy usage analysis and supplier cost comparison.

Set achievable objectives and make it clear that they are relevant to the purpose of the plan and that they are in line with the financial projections and other aspects of the environmental plan.

Audit of environmental management objectives, policies and activities

Periodically, an organization should conduct a complete review of environmental management objectives, policies and activities on an organization-wide basis. Such a review is a comprehensive approach to evaluating environmental management effectiveness. The review aims to examine and evaluate the success or otherwise of the environmental management objectives and policies which have been guiding the organization. This is a comprehensive review of both the activities of the organization in relation to environmental management, and also the environmental factors that are likely to contribute to success, or failure, in achieving the objectives of the environmental management plan.

Environmental management plan budget

The environmental management plan budget provides the foundation on which key strategic and funding decisions are based. When writing the financial element of the plan, detail the environmental management budget and the ways in which it relates to the targets that were developed earlier in the environmental management plan. It should include a statement of the finance required, its intended purpose and the projected impact on the profitability of the business.

Budgets and resources are allocated in an attempt to meet business and environmental management objectives and to make the overall environmental management plan successful. Products performing particularly well or particularly poorly, from an environmental perspective, obviously require quite different levels of activity, and consequently, differential allocation of budgetary resources.

Often, the costs involved in the environmental management plan are a very significant part of the overall operating costs of an organization. It is sensible to plan these costs systematically, to attempt to forecast their effect, and to keep as close control as is possible over the way in which they are used.

Keep the budget section brief, but comprehensive. The environmental management budget can be as comprehensive as the financial plan that would be written as part of a business plan. Many of the items may be identical, and could be transferred directly from the business plan or profit and loss forecast.

The following represent key tasks to be undertaken to develop an environmental management plan budget:

- Estimating total revenue and costs.
- Allocating environmental management resources within the portfolio of products/services and between elements of the environmental management objectives.

Allocation of the environmental management budget

The environmental management budget has to be allocated to the various departments or elements of the environmental management plan. New products, or products that the organization is attempting to reposition environmentally, may require a higher environmental management budget commitment than others. Estimates of additional revenue for new environmentally friendly products or by-products should help to make the allocation decisions a little easier.

3

ORGANIZATIONAL BARRIERS TO ENVIRONMENTAL MANAGEMENT SYSTEMS

Summary

Within an organization there are many strategic and operational activities that can act against the successful implementation of the most basic of EMSs. This chapter identifies these barriers and discusses each of them to determine the extent to which organizational barriers impede the implementation of an EMS. The identification of these key organizational barriers may assist managers to make informed decisions and allocate scare resources efficiently and effectively, particularly if those barriers that facilitate the implementation of an EMS could be determined from those that act against it. Growing public pressure on organizations to adopt production systems that do not unduly impinge upon the natural environment is reflected in the adoption of EMSs in business activities. An EMS can change an organization's structure, responsibilities, practices, procedures, processes and resources for environmental management, so that it is able to reduce negative environmental impact while improving management control (Renwick *et al.*, 2008; Bansal and Hunter, 2003). In addition, a certified EMS such as ISO 14001, a globally recognized standard for environmental management, provides a strong signal to external stakeholders of its environmental management commitment (Linnenluecke and Griffiths, 2010; Jiang and Bansal, 2003).

Therefore, it is expected that EMS implementation brings about organizational benefits with an increase in the demand of environmentally conscious customers, achievement of environmental objectives and cost reductions through improved productivity. According to Segersen and Miceli (1998) and Welch *et al.* (2002), it is widely believed that organizations adopt such environmental standards because they recognize that the accrued benefits, not just production and economic benefits (Jabbour *et al.*, 2008) of doing nothing. Although the effect of EMS implementation through an increase in demand is a direct effect, that through an improvement in productivity is regarded as an indirect effect, because an EMS only provides a management framework for the environmental objectives and it is expected that several environmental management activities for the objectives actually improve productivity and provide an economic benefit (Hertin *et al.*, 2008).

Models

The concept of SD is based on three major aspects of organizational performance: economic, social and environmental performance. However, defining environmental performance is no easy task. Environmental performance may be defined using a two-way matrix (see Table 3.1). The vertical axis shows the process and outcome dimensions while the horizontal axis reflects the internal and external dimensions. The junction of these two axes determines four dimensions of environmental performance:

- Enhanced products and processes – competitive advantages obtained by an organization as a result of its environmental initiatives.
- Relationships with interested parties – the interaction between an organization and its various outside stakeholders, including shareholders, the local community, government, clients and suppliers.
- Regulatory compliance and financial impacts – the level of response to environmental standards required by laws and regulations, as well as the economic consequences of environmental initiatives.
- Environmental impacts and corporate image – the negative externalities of an organization's activities on its environment and its overall reputation. An organization's environmental performance can be measured in concrete terms using a variety of indicators.

Generally, environmental performance indicators (EPIs) may be defined as measures that are based on observable or determinable quantities that reflect in

TABLE 3.1 Elements of an organization that can impact on the implementation of an EMS

Elements of an Organization	Includes
The structure of the organization	Departments, subsidiaries, management hierarchy
The environment in which it operates	Geography, economic climate, legislative frameworks
The decision-making process, or management of the organization	Democratic or authoritative management styles, management hierarchy
The people within the organization	Personal beliefs, priorities, level of education
The general way in which change is viewed and implemented	Management style, management hierarchy, organizational culture (proactive, reactive or passive)

various ways the environmental impacts of a given activity. The quantities involved may be physical quantities of materials used in an industrial process such as energy, water and raw materials, or that result from the process (emissions into the environment in the form of air pollution or liquid effluent). EPIs can also be used to measure the efforts made to reduce impacts, for instance spending on environmental management, or implementing an environmental management system. These indicators may also reflect several perspectives: economic versus ecological, input versus output, process versus outcome, internal versus external or operational versus managerial. Among other things, indicators make it possible to identify trends, causal relationships or progress in set objectives. They can be useful when responding objectively to enquiries about the status of an organization's environmental performance.

There are a number of different models to guide the development of EPIs. Among the best known are those of the International Organization for Standardization (ISO) and the Global Reporting Initiative (GRI). ISO standard 14031, which belongs to the ISO 14000 family of standards, proposes three types of EPI: environmental condition indicators (ECIs), operational performance indicators (OPIs) and management performance indicators (MPIs). ISO standard 14031 also provides a framework for the selection of the most appropriate EPIs, along with tools to identify and integrate the aspirations of interested parties. Several practical examples of its application are included in the various guidelines.

The performance indicators provided by the GRI extend to all dimensions of SD, that is economic, environmental and social performance (Coleman, 2005) and add a fourth dimension related to integrated performance. Basically, the latter is designed to unify all performance dimensions and support deployment of an organizational strategy. With respect to EPIs, the GRI suggests 35 types of indicators, including 16 core indicators that can be used to measure all aspects of environmental performance (materials, energy, water, biodiversity, emissions, suppliers, products and services, compliance, transportation and overall performance).

Measuring environmental performance through an index

Too many measures can make information unusable. Some indicators are very useful for day-to-day management, but ineffective for strategic planning. To simplify the assessment of an organization's overall environmental performance, EPIs can be aggregated into a single index. By allocating different weights to each value in the index, a score can be determined to assess an organization's overall performance. This makes it possible to identify, at a glance, the direct impact of several series of activities for a given project and to define their volume and importance. Using an index allows current multidimensional measuring systems to be compared by synthesizing the information of each system into a single value or in the form of a diagram. Should it be appropriate to get to the bottom of unexpected changes, the index can be examined to ascertain where the largest changes occurred and to make the relevant corrections.

However, aggregating several EPIs into a single index isn't an easy task. Since the main purpose of environmental performance measurement is to convert a large quantity of data into information that is useful for management decision-making, it's crucial to weight the items that make up the index. Positioning targets for a combination of indicators that perfectly reflect a range of objectives is difficult, if not impossible. However, defining objectives for critical resource indicators is extremely challenging. Indeed, some critical thresholds and limits are regulated for certain emissions, but for most of them, the interconnections of environmental factors make accurate assessments difficult.

Indices should provide a balanced vision between resource use and the quantity of manufactured goods and services, emissions, effluents and waste of all kinds to assess the effectiveness and eco-efficiency of a business process. However, development of indices always involves a measure of subjectivity associated with the values and priorities of the individuals designing them. One way of getting around this problem is to implement an open and transparent process for the development and assessment of EPIs by involving interested stakeholders in the community and environmental impact assessment process. This is pertinent to promoting sound corporate responsibility, environmental management and continuous improvement.

Case study: Norsk Hydro Canada

Since 1987, Norsk Hydro Canada has been manufacturing primary magnesium in its facility at Becancour, Quebec. Along with its ISO 9001 certification, this organization has the world's first magnesium plant certified ISO 14001, and the first production plant ever to recycle magnesium. This facility is also recognized as a leader in environmental management. In fact, it has received several high profile awards in this field, including the EcoGESte 2001 award for best performance in reducing greenhouse gas emissions over the past 10 years, and the Énergia 2005 award for sustainable initiatives in the institutional sector, which recognizes excellence and merit in achieving energy efficiency and control.

Norsk Hydro's commitment to eco-efficiency is reflected in several of its achievements, including a $1.5 million per year reduction in brine and acid losses and the implementation of an R&D programme to replace sulphur hexafluoride (SF_6) completely. This programme is in full swing, as Norsk Hydro has been using a new gas mixture since the end of 2004. The use of this new gas has reduced consumption of SF_6 by more than 90 per cent, in addition to generating a 900,000-tonne reduction in carbon dioxide equivalent from 2004 to 2005.

An organization's environmental management philosophy is based on three corporate commitments:

- Compliance.
- Risk reduction.
- Resource conservation.

Compliance hinges on the observance of environmental regulations prescribed by various levels of government, the performance indicator used represent the number of times that a standard is exceeded during a month. Risk reduction is primarily concerned with preventing and eliminating any environmental incident and to protect employees and the general public in the event of an occurrence. Crucial to risk reduction is the response time required in the event of an occurrence. There exists a direct relationship between response time and the costs generated by an environmental incident. Accordingly, the organization has emphasized training employees so that they can detect and react immediately as situations warrant. For instance, plugging a leak at source may cost hundreds of dollars whereas if the leak reached the drainage system it could cost many thousands of dollars.

The organization's final commitment, resource conservation, reflects a desire to reduce losses of raw materials and by-products, as well as polluting emissions, at source. The Chief Executive stresses that, at Norsk Hydro, pollution is synonymous with loss of resources and money. He also notes that 80 per cent of environmental costs are due to wasted resources, regardless of their form, which is why the organization implemented its FEEW environmental performance index, for the loss of raw materials in effluents (F), emissions (E), energy (E) and waste (W). This initiative saved $4,000 per day.

The cost of resources purchased or produced was easily identifiable. All that remained was to measure losses, using various techniques. By multiplying the quantity of resources lost to air, on the ground, in the water and in energy by their cost, Norsk Hydro has managed to quantify the cost of inefficiencies on an annual basis. Thus, the organization is able to assess its environmental and economic performance year after year, in addition to ensuring quality control at the operational end.

The entire process is based on the three Cs – reducing what can be comprehended, calculated and controlled – as well as on the services of people who know, want and can (KWC). Those who know – the environmental team – are responsible for establishing priorities and highlighting inefficiencies. They make their decisions on the basis of a 'reduce, reuse, recycle and upgrade' mantra.

At the next stage, those who know call on those who want to reduce inputs and improve production processes. As more productivity gains are generated, more executives are willing to invest in making life easier for 'those who know and can'. Finally, the organization involves 'those who can' act directly on processes and the choice of equipment – engineers and operators. The initial focus is on the low-hanging fruit – the quick wins. All stakeholders are motivated by success, and the most significant savings are often apparent early on. Letting the entire organization know about the environmental and economic gains achieved is crucial to keeping everyone focused and motivated. At Norsk Hydro, these results are communicated through a bulletin and monthly reports to people who are directly affected by the results such as directors, stewards and executives.

Using environmental performance indicators

The Norsk Hydro example demonstrates how EPIs can be used to control costs and production processes, while alleviating environmental pressure and maintaining a motivated staff. However, an EHS can be utilized for implementing strategy. Without reliable data, even the most well-intentioned executives can't identify zones of excellence and areas for improvement. To be truly meaningful, EPIs must be linked with clear organizational goals and objectives. These indicators track and disseminate specific differences from established objectives, and the resulting trends are then analysed to promote continuous improvement. This also fosters organizational communication, since such indicators represent signals that are carried through the organization. Furthermore, because they provide feedback, EPIs also contribute to organizational learning. Disclosure is greatly enhanced through EPIs, as these data can be directly integrated into annual and SD reports. Without them, a quantitative assessment of environmental impact reduction would be impossible.

Beyond external disclosure, an organization can also use the various EPIs to direct its decision-making. Making decisions involves choices such as assessing energy efficient capital projects, selecting vendors who are more respectful of the environment, and planning production based on the environmental impact of the products. At a time when more and more businesses are contemplating environmental information that goes beyond traditional financial and operational efficiency data, very few of them relate this information to the compensation of the people in charge of their implementation. The same goes for using compensation to undertake organizational change. It's recommended that EPIs be tied to financial incentives and used to undertake and assess strategic change. Otherwise, it may be more difficult to integrate EPIs into the organization's guiding and managing processes.

Economic versus environmental development

Although the majority of scientists agree that climate change and global warming are man-made phenomena and pose serious economic, environmental and social threats to future generations, society struggles to fully engage with this message. In fact, with each severe winter, or cool summer, society's voice is heard challenging the existence of such phenomena. Putting the climate change, global warming debate to one side for the moment, the focus turns to global economic challenges, particularly ever increasing energy costs and security of supply. Economies are reliant upon fossil fuel energy sources and focused on meeting the increasing economic and lifestyle demands of every citizen, so each economy moves closer to economic crises with ever increasing energy costs, lack of energy security and depletion of natural resources. Lobbyists of industry sectors reliant upon fossil fuels advise that resources may last another 20 or possibly 40 years. There is no doubt that fossil fuels, at current rates of consumption will run out very quickly, if as suggested 'tipping points' have already been passed (Porritt, 2005). The uncertainty lies in the definition of 'very quickly'. What is certain, however, is that in terms of the lifetime stages of an industry, those reliant upon fossil fuels as energy sources are in an 'end game' scenario. That is, as demand grows and energy resources become increasingly scarce so prices increase and inhibit sustainable economic growth.

The overriding view, at this time, particularly looking at national policies and strategic direction on how to overcome the twin challenges of climate change and energy security, is rapid development and diffusion of low carbon technology. Certainly, innovation and technology development played a leading role in the realization of the industrial revolution and the more recent communication and digital revolutions. It could be argued that these technological revolutions, and therefore technology itself, have been the cause of many environmental concerns today. However, technology is, for better or worse, out of Pandora's Box; returning to a less technological way of life is not an option. Innovation and technological development by their very nature will progress with increasing speed. Despite this push for technology, caution should be taken of the concept of 'technology lock in' and 'generation technology'. Historically, technological development has had, predominantly, an economic imperative. It is hoped, whether by accident or design, that the development of low carbon technology mitigates the impacts of a high carbon economy and assists the movement towards a low carbon economy, a new society where the economic driver is synchronized with social and environmental needs.

Political and economic context

In Europe, the EU political agenda seemed to change dramatically within a very short period of time. New priority objectives now include a low-carbon and resource-efficient economy and developing a sustainable industrial policy. The pursuit of low carbon technology development is acknowledged, by many European governments, as fundamental to the delivery of these objectives. EU heads of government are committed to tackle these challenges (European Council, 2007). Ambitious targets for GHG reduction, energy efficiency and renewable energy production have been set and EU leaders agree that low carbon technology development is one of the key pillars to building the future EU economy. The EU economic competitiveness is, to a large extent, based on its energy and resource efficiency, and its ability to develop appropriate technological solutions to environmental problems. Energy efficiency, as well as resource and material efficiency, is central to both economic growth and environmental protection. Achieving these objectives requires reflection upon how to better use EU funding instruments to provide targeted stimuli to finance technology development and innovation, as well as the take-up of low carbon goods and services. In Northern and Western European countries, technology leadership and economic growth play crucial roles in the structure and deployment of funding instruments. Further goals of financial institutions are to create new jobs and encourage environmental improvement, and return on investment also plays a more important role in Northern and Western European countries.

The transition to a low carbon economy and the deployment of resource-efficient technology requires a range of custom-designed financial instruments to suit a range of businesses, at various stages of technological development. This approach needs to be supported by a combination of private and public sector capital and resources to be successful. The existing lead market initiative recognizes that public institutions require commercial funding support on a large scale. Investment flows need to transcend existing public–private practices in order to meet this low carbon technology financing challenge.

It is, therefore, not possible to predict what a precise path to a low carbon economy entails, either in terms of technologies or policies to pursue as this varies from country to country. Such a scenario would suggest that each country should follow a path of emissions reduction, energy security and economic growth that best suits its stage of development, available resources and its culture. It should also identify whether the economic cost of following such a path is feasible, given the range of low carbon technologies and economic growth policy levers which could be deployed.

Public sector low carbon technology funding

Securing future energy supply in a sustainable manner and protecting the environment are some of the most important stakeholders in current policy-making processes. Since the costs for renewable energy technologies have decreased significantly over the last 10 to 15 years due to technology improvements, institutional learning and economies of scale in production, and public awareness of climate issues has increased considerably, alternative energy sources are seen as a viable option.

Public sector intervention

Effective government intervention promotes low carbon technology innovation and provides the operating frameworks that financiers need to take long-term investment decisions. Interventions assist in creating markets, reducing risk, providing acceptable rates of return for investments and creating conditions for a sustainable and profitable industrial sector. Therefore, a long-term regulatory framework plus finance mechanisms that address barriers and gaps to financing sustainable energy as well as low carbon technology are necessary tools for the development of this sector.

Public sector financing mechanisms

Supportive national regulatory and tax environments are key essentials in promoting economic development and financing conditions for new technology development. Differences between conventional and sustainable energy projects that require a need for public intervention concern factors such as financial scale, capacity, energy resource characteristics and status of technology, which consequently require new thinking and new risk-management approaches as well as new forms of capital. To create a real enabling environment needed for large-scale investment in low carbon technology, public finance mechanisms are required to fill funding gaps and address barriers to technology development as it proceeds from research stage through to commercialization and full-scale deployment. These barriers and gaps may vary, depending on the applicable economic and regulatory environment, regional contexts, stages of technological development and current market trends.

Funding goals and stages

A strong overall goal of funding low carbon and sustainable energy technologies within the financial instruments is with the environmental improvement in general. However, the second important target is economic development. If other economy related goals and targets such as business start-up, employment growth and industrial development are being subsumed under the heading of economic development, then this goal becomes most important. That can be interpreted as a political strategy to create a win–win situation and to bring together environmental improvement and low carbon economic development.

Organizational barriers

The general principles that apply to any organization offer a good basis on which to study the influence of environmental management within organizations. There are some obvious 'hard' technical barriers and benefits to EMS within the specific operations of a organization, such as equipment, new technology and innovations, but there are also some key barriers to EMS implementation based within the 'softer' non-technical aspects of an organization such as organizational structure and culture (Stone, 2000; Ramus, 2002).

It is incumbent upon management to investigate emerging organizational responses to EMS and the extent to which the responses facilitate or impede the implementation of an environmental management system. The identification of key organizational barriers may assist management in allocating scarce resources more effectively. These resources can be allocated more efficiently and effectively if those barriers that facilitate the implementation of an EMS could be determined from those that may be acting against it.

The identification of barriers to EMS implementation is necessary to enable corporate corrective counter measures. However, corporate managers are finding themselves with an increasing number of barriers with which to counter if EMSs are to be introduced successfully. Tackling all of the identified barriers may be considered an inefficient use of resources, particularly if some barriers have a greater impact than others. Managers may welcome a more analytical approach that endeavours to identify key barriers on which scarce resources can be focused to achieve the greatest benefits.

Corporations at the early stages of implementing an EMS could be aided by the identification and measurement of those barriers that are likely to emerge during the process. Such measurement would allow corrective action to be taken with greater precision and ensure a more efficient use of resources. For the long term sustainability of an organization an EMS adds to the traditional financial (economic) imperative by also managing for environmental and social issues, hence a reliance on environmental performance for corporate sustainability (Jacobs *et al.*, 2010).

Environmental issues, particularly environmental legislation, affect all sizes of organizations and some organizations are more active than others in addressing these issues (Hunt and Auster, 1990). What is clear from the literature is that larger organizations are more likely to have EMSs than small to medium-sized enterprises (SMEs) due mainly to the availability of additional resources (Welford, 1996). They are also more likely to have formalized structures and a dedicated environmental manager to introduce and monitor their EMS. As a consequence the number of organizational barriers that exist within large organizations is likely to be high and their interaction more complex.

The existing environmental management and organizational theory literature identifies 12 main organizational factors that could potentially act as barriers to the implementation of an EMS (Tinsley and Pillai, 2006). These potential barriers are detailed in Figure 3.1. The following descriptions of the barriers serve two functions. First, to support its inclusion as a potential barrier and second, to identify the characteristics of each barrier, generate a clearer definitions which both defines it and differentiates it from the others (Eisenhardt, 1989).

Available resources

The introduction of an EMS may be hampered by the shortage of adequate resources or by the lack of recognition or provision of necessary resources (Greeno and Robinson, 1992). The lack of available budgets, human resources, and corporate incentives (Tapon and Sarabaru, 1995; Gallarotti, 1995) are identified as potential barriers. A lack of available resources or the misallocation of resources may result from other existing barriers such as a lack of commitment or lack of communication (Kirkland and Thompson, 1999). Additionally, management decisions about resource allocation that forms the basis of implementing an EMS are not made in isolation. The decisions would also consider the realization of economic benefits, as well as the human resource issues of increased job satisfaction, staff retention and the recruitment of future talented staff (Branco and Rodrigues, 2009; Jabbour *et al.*, 2008). A recent study by Wagner (2013) identifies the positive benefits of employee satisfaction and staff recruitment and retention that arise for human resource management with the introduction of a well-resourced EMS.

Communications

A lack of internal and external communication can act as barriers, particularly if not diffused or cascaded successfully throughout the organization (Fuller and Swanson,

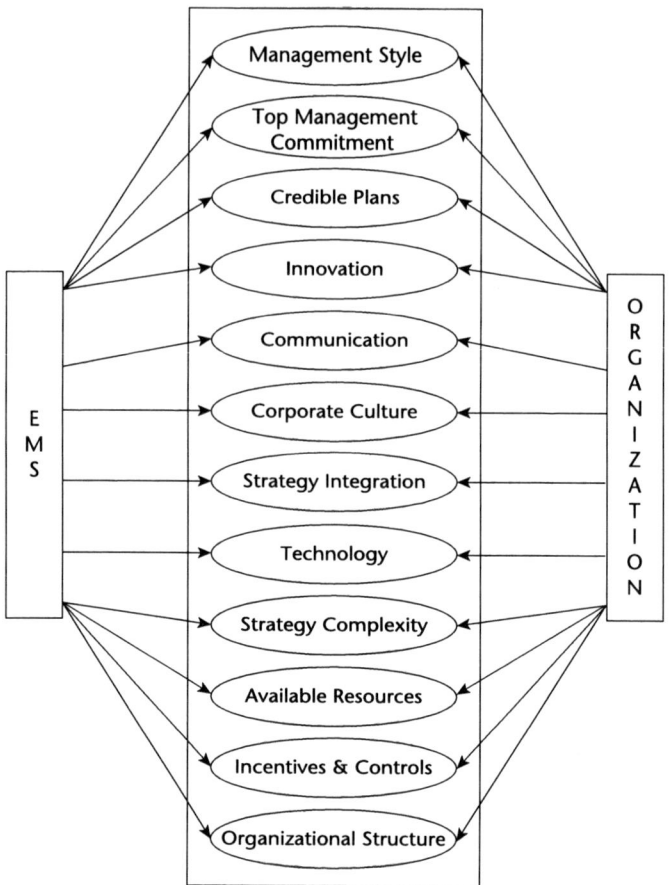

FIGURE 3.1   Organizational barriers

1992; Azzone and Bertele, 1994). While a number of organizations are attempting to introduce EMSs, those managers tasked with the responsibility of system implementation often do not receive the necessary training and support (Kirkland and Thompson, 1999). Those managers who have the required levels of training and support know the importance of raising levels of environmental knowledge among all employees.

With change, the existing behaviour patterns and values of an organization are often questioned. Individuals can feel threatened particularly if there is a lack of communication to aid understanding of the new organizational direction and the new roles of its employees. The use of departmental briefings, organizational seminars, work groups, intranet and newsletters are useful formal methods of communicating change. Informal communication methods such as 'corridor briefings' or 'canteen chat' can also be useful.

The best option can often be a mix of both as long as a consistent message is being communicated. Whatever communication methods are used, if change is to be managed successfully then the communication process needs to be two-way. Senior management must not be perceived to have all the answers and delivering directives to middle managers. Presenting the problem to middle managers to solve can be very rewarding, not just for the managers but also for the successful implementation of the new EMS.

Another way in which communication can be a barrier to an organization is if it does not cultivate every possible means to communicate with their customers and stakeholders. Utilizing a range of periodic briefings such as business breakfast events, newsletters, published reports, website logs and other social media outlets help customers and stakeholders know of an organization's environmental commitment and credentials. The implementation of an EMS is being accepted as an image-builder to strengthen the organization's competitive position (Hui *et al.*, 2001), but this competitive advantage can only be realized if customer and stakeholder communication is effective.

In a study undertaken to explore environmental communication in organizations (Ramus, 1998), key examples were identified as to how communication could be improved:

- Continuous reiteration of the organization's overall environmental vision by the senior managers.
- Involvement of all staff in environmental projects and programmes, internal and external.
- Annual updating of environmental policies, manuals and environmental reports, which should be distributed to all staff.
- Adopting both top-down and bottom-up communication of environmental information through meetings, electronic media, and suggestion boxes.
- Employee surveys to assess the effectiveness of the environmental communication system.
- Accreditation certificates and awards campaigns to keep employees aware of goals and target achievement.

Corporate culture

Organizations that are similar in many ways respond in often radically different ways when it comes to environmental management and practices. An organization's culture is one explanation for the difference in its behaviour. The staff and other stakeholders of an organization expect the rules, procedures and processes of environmental management to reflect the values and beliefs of an organization. And this organizational culture drives the actions and behaviour of staff (Berry, 2004).

An organization's culture can be defined by the prevailing values and attitudes acting within it that have traditionally been relied on or previously adopted (Welford and Gouldson, 1993). Due to the speed, range and complexity of environmental issues, organizations need to internalize and operationalize policies and programmes to be consistent with long-term goals (Corbett and Wassenhove, 1993). Culture change produced long-term business benefits to organizations such as Total. Previous, 'green-wash' or public relations management of environmental issues has proved to be ineffective and has, at times, back-fired on organizations, damaging reputations (Peattie, 1990).

Organizational culture reflects the way people perform tasks, set objectives, administer resources and the way they feel and respond to daily opportunities and threats. Culture is so fundamental to the way an organization works that people respond in an unconscious manner to 'a way' of working within it. The prevailing culture within an organization is the driving force for the whole organization.

Stone (2000) listed elements of an organization's culture that are relevant to the uptake of cleaner technologies and identified the importance of the human dimension to the successful implementation of environmental improvement. Three key elements are discussed in more detail below:

- Commitment of decision-makers and style of management.
- Encouragement and training of staff.
- Strategic attitude.

Commitment of decision-makers and style of management

Both commitment of senior management and management style have been discussed separately, but both are key to the culture of the organization. Whatever style leads the organization, the culture is created through the initial setting of objectives. The setting of policies, structure, control and reward systems to achieve the objectives also reflect the culture (Mintzberg, 1987).

Encouragement and training of staff

Organizations that show trust in their employees to act responsibly have successes in implementing a culture that demonstrates environmental improvements. Employees should be encouraged to take ownership of the processes that they are involved it but also look outside of their own job responsibilities (Ramus, 2002). It has been stated that competitive advantage derived from environmental improvement is found in those organizations that empower staff by making use of their ability to learn and give them the tools with which to apply this learning to all areas of the business (Hutchinson, 1996). A lack of specialist knowledge due to a lack of provision for training outside the remit of the main job is also a potential barrier to reaching environmental objectives.

Environmental training may be incorporated into all levels of existing training programmes within an organization, or specific environmental programmes may be offered. Some organizations offer external environmental educational opportunities such as job rotation and site visits. However, supervisory encouragement still remains a key factor in the success of any type of employee training (Ramus, 2002).

Strategic attitude

A key part of an organization's culture that can impact on the successful implementation of an EMS is its strategic attitude, also referred to as management approaches (Azzone *et al.*, 1997). Three key strategies are highlighted to enable an organization to position itself to competitive challenges:

- Compliance-based attitude: an organization takes action only according to external pressure from competitors or the wider market.
- Anticipatory attitude: while reacting to external pressures, an organization anticipates future changes and implements proactive initiatives.
- Innovative attitudes: an organization implements changes in management procedures to introduce a new product to an existing market, therefore creating a breakthrough in product performance.

Clearly from these categories, the more passive approaches are going to lead to a less proactive pursuit of an EMS. Ironically, a compliance-based attitude could work more effectively if an EMS was in place, as it could react rapidly to external pressure as and when the need required.

For the two more active attitude categories, which are therefore more likely to implement EMS, the structured framework of an EMS aids an anticipatory attitude to recognize future change, and the innovative attitude enables speedy identification of key procedures or processes that are opportunities for innovation.

Other authors have categorized management approaches in similar ways, all based on the proactive stance of the attitude. Vastag *et al.* (1996) list four approaches: Reactive, Proactive, Strategic and Crisis Prevention. The categorization of organizations according to their attitude and circumstances is a useful tool for academics to study effectiveness of EMS. These categorization models are discussed in more detail in Chapter 4.

Credible plans

Inappropriate or 'quick fix' plans to introduce an EMS into an organization are a risky strategy (Kirkland and Thompson, 1999). The use of credible, well formulated plans can benefit the organization in three main areas. First, environmental issues can pose complex problems for organizations; therefore the plans created must highlight and address the complexity by reducing it to its simplest form. Second, they reduce the potential for management mistakes that can quickly lead to disenchantment and loss of credibility and create general resistance to the introduction of the EMS. Finally, the sustainability of an organization has traditionally been measured in financial terms, and today, corporate sustainability is also measured socially and environmentally, hence the need for an efficient and effective EMS.

Good planning that is based on good analysis of good data can have many EMS benefits such as:

- Increasing general awareness.
- Reducing risk in decision-making.
- Identifying 'aspects' and 'impacts' of products, processes and services.
- Establishing agreement on organizational priorities.

The key difficulty in formulating credible plans is ensuring that associated policy statements, procedures, plans and budgets are coordinated.

Incentives and controls

It is difficult to introduce EMSs and programmes into an organization through senior management directives, as there is a need for incentives and controls to be in place to better ensure staff support. Pascual and Gomez-Mejia (2009) suggest that when managers are rewarded for good corporate environmental improvement this acts as a good signalling mechanism for staff and improves work satisfaction levels. Those effective, environmentally aware organizations ensure that managers and directors agree to deliver environmental objectives together with budgets as part of any reward schemes to achieve corporate environmental objectives (Tinsley, 2002).

Innovation

Such factors as consumer demand, government influence and legislation influence the rate, scale and type of industrial innovation within organizations (Foster and Green, 2000). As a consequence, the development of a new or even modified environmental product can take up a lot of time and scarce resources before it becomes part of an organization's product portfolio.

An organization's decision-making process for the acceptance of promising environmental innovation may also be biased. In exploring the use of innovation in environmental management improvement, Cramer and Zegveld (1991) found that innovative solutions eventually selected appeared to be the most successful, or advantageous, in a competitive, as opposed to an environmental improvement sense.

Traditionally, innovation has been identified as causing delays, obstruction, misunderstandings and disagreements in newly installed organizational systems (Irwin et al., 1994; Roome, 1994). Innovation, or the lack of it, was also seen as an issue in the implementation of EMSs. Porter and van der Linde's (1995) study suggests that too many organizations spent too many environmental pounds on fighting regulation and stalling legislation when they should have been looking for real environmental solutions to environmental issues.

Today, governments, institutions and large organizations, particularly, see innovation as the answer to achieving healthier economies and a competitive advantage for economies.

Rennings *et al.* (2006) suggest that for most organizations, development divides innovation into three types:

- Process innovations enable the production of greater output with the same or lesser input.
- Product innovation improves goods or services or leads to the development of new products.
- Organizational innovation refers to new forms of management.

Environmental innovation is then divided into two types:

- Innovations in end-of-pipe technologies: usually add-on measures that allow a piece of equipment to comply with legislation such as filters, sound absorbers and sewage treatments.
- Innovation in integrated technologies: modifications or new equipment for cleaner production to remove the harmful by-products at source throughout the process.

The process of innovation can be undertaken with small steps or giant strides, but whatever method is selected, barriers can arise. A slow, incremental, or small steps approach can signal low importance to staff on the corporate priority list and provide competitors with time to deliver similar or alternative changes to rival products or services. A radical approach can create a force-field of resistance through competition for resources, lack of adequate planning and lack of understanding of product marketing and manufacturing requirements. The whole society, potentially, benefits from new environmental innovation; however, it is the organization that has to meet the costs, much of which cannot always be recouped through effective marketing, particularly if competitors are able to respond quickly by copying the improvements (Rennings *et al.*, 2006).

Manager and employee role in innovation

Employee creativity has a role to play in innovation, as it is an important problem-solving resource in an organization. However, in order to use this resource, organizations must provide the systems with which to support employee action, and managers often do not give the same level of support to employee environmental activities as general management tasks (Ramus, 2002).

The study conducted by Ramus (2002) identified that it is the role of environmental policy to show employees that the organization desires environmental ideas or actions, but that this needs to be supported by direct encouragement from supervisors and managers.

Ramus (2002) has split employee eco-innovations into three types:

* Those that decrease the environmental impacts of the organization.
* Those that solve a particular environmental problem for the organization.
* Those that develop a more eco-efficient product or service.

Innovation refers to the creation and successful exploitation of new ideas. It includes the development of new technologies, products and processes, the opening of new markets and more intangible activities such as new methods of working and organization. The innovation process covers all stages of bringing new products and processes to market from basic and applied research through to early stage development and pre-commercialization demonstration.

Innovation has the potential to provide benefits to individuals, organizations and the economy as a whole. Consumers benefit from better value goods and services and more choice, which provides greater opportunities to find products that better suit their needs. Innovation is also a fundamental element of business competitiveness and of public sector efficiency. Organizations are encouraged to invest in innovative activity to exploit opportunities for cost reduction and enhanced profit. The incentive to innovate is also strong where organizations are faced with challenges that could threaten their future performance and profitability.

The transition to a low carbon economy provides many challenges and opportunities to advance the development of technologies. Businesses depend on innovation to help them improve the efficiency of their existing processes or justify investments in new low carbon technological processes. Those that innovate successfully gain first mover advantage and improve their competitive position to gain market share. However, existing business development and marketing management literature suggests that new products introduced to create new markets are a high risk strategy that has a low success rate; less than 1 in 3.

There are several sources of innovation. The more 'traditional' view of innovation places the activity internal to the organization, for example manufacturer innovation – focusing on internal research and development. This is essentially a 'closed' and linear model of innovation whereby the organization retains control of development, manufacture, marketing and distribution. However, there is increasing recognition and application of a more 'open' model of innovation, an end-user innovation where economic agents develop their own products or processes where the market is unable or unwilling to generate something that sufficiently meets their needs. This activity has been identified by some as a fundamental part of the innovation system, and in some cases potentially more important than conventional 'closed' innovation routes.

A simple distinction is drawn by Swann (2009) between innovation that is incremental and those that are considered radical. Such a classification follows those defined earlier by Kline and Rosenberg (1986), Schumpeter (1942) and Freeman and Soete (1997) among others. Other classifications exist within this broad spectrum; however, regardless of the classification used, all forms of innovation have the potential to provide, through facilitating productivity improvements, more efficient ways of using resources and to generate the means to reduce the environmental impact of economic and social activities.

Incremental innovation refers to the steady stream of improvements to an established product, process or technology which does not change its fundamental character. Each individual change may be relatively minor, but the cumulative impact of a continuous flow of incremental improvements can, if sustained, have a significant cumulative effective on productivity. A large amount of innovative activity is incremental in nature, manifesting itself through a number of different routes including open source innovation and adjustments to processes 'on the factory floor' as well as more traditional routes of research and development activity.

Incremental innovation

Herbig (1994) highlights three types of innovation within an incremental innovation process:

- Continuous innovation, the introduction of a modified product.
- Modified innovation, a slightly more disruptive innovation, for example the introduction of a new technology that performs the same or very similar basic functions.
- Continuous improvement, improvements in the way existing products are being produced, for example, through the introduction of TQM and systems re-engineering.

Incremental innovation is a key part of the push towards lower carbon emissions, primarily through the need to improve the energy and wider resource efficiency of existing technologies. This might involve innovations that improve conversion ratios in energy generation, increase the fuel efficiency of vehicle engines and improve the use of natural resources in production.

Radical innovation

Radical innovation, on the other hand, refers to that which significantly alters the overall technological profile of the economic system. Innovations of this type have the potential to completely undermine the position of market incumbents, whereas incremental innovations are less likely to do so. Schumpeter (1942) outlined this idea of 'creative destruction' by suggesting that innovative entry by entrepreneurs was the driver of sustained long-term economic growth, even though this activity may destroy value of market incumbents who enjoy some degree of market power. For example, renewable energy, electric cars.

In order to overcome potentially significant barriers to entry that protect incumbents from competitive pressure from similar products, new entrants would have to be radically different, leading to innovative activity that leads to a fundamental improvement. In addition, the threat of competition from radically different products and new market entry puts pressure on incumbents to remain competitive and undertake innovative activity themselves. At the outset, radically innovative technologies may appear highly speculative and high risk, although once proven they become fundamentally embedded into the mainstream activity in an economy and can often require complementary and major infrastructural change. Helpman (1998) refers to these types of innovation as 'general purpose technologies' that have the potential for pervasive use in a wide range of sectors in ways that significantly change their operation. Examples of such major breakthrough technologies include the development of steam power, electrical technologies, and, particularly important for the structure of the modern economy, the introduction of computers.

In the context of moving to a carbon constrained economy, low-carbon technologies may be direct substitutes for high carbon equivalents, representing a significant shift in the technological base, even though the output is relatively similar. For example, consider renewable energy technology replacing fossil-fuel intensive energy generation, or electric, biofuel or hydrogen-fuelled vehicles taking over from mainstream petrol or diesel ones. In some cases, these 'transformative' innovations may require substantial infrastructure investments reflecting the widespread implications of the development of these technologies. For example, the development and widespread introduction of electric-powered vehicles requires an extensive supporting infrastructure to ensure sufficient access to charge points. Similarly, the development and application of carbon capture and storage technology (pre or post-combustion) in fossil-fuel powered generating plant requires an extensive grid for CO_2 transport and storage. Offshore wind farms require similar support infrastructure similar to that of the existing oil and gas industry.

The divisions between these different modes of innovation are not clear cut – it should not be assumed that incremental innovations only represent 'small' changes and radical innovations refer to 'large' shifts. To some extent the label attached to each innovation depends on its disruptive impact on the rest of the market and the wider economy. It should also not be assumed that lower levels of effort are required to achieve incremental improvement while only large investment leads to radical change. Large levels of investment could be required for some small but vital incremental improvements whereas in some cases minimal investment can lead to significant changes. The distinction therefore is in the impact of the outcome, and not the size of the input. The difficulty of trying to gauge the level of input or effort required in advance means that innovation is a fundamentally high-risk activity.

Management style

Change within an organization occurs as a result of signals received from the external environment, and the interpretation of these signals by management results in decisions being made and actions taken that change the internal aspects of an organization (Mintzberg, 1987). Different managers interpret these external signals in different ways and make business decisions that suit their personality, experience and career objectives. Also, Ralston *et al.,* (1999) identify that the attitudes of management change with each new generation, and point to the increased awareness in younger managers of preserving the natural environment and improving corporate responsibility.

An organization that succeeds in achieving growth and innovation in environmental performance is often one that has managers who are good at empowering employees. Management support is often more important to staff than an organization's written commitment to environment performance, as supportive behaviour from someone in a supervisory role is seen as direct encouragement (Ramus, 2002). Management style incorporates a number of factors:

- Risk taking.
- Levels of delegation and planning.
- Teamwork.
- The ability to change as new opportunities and threats present themselves.

The various leadership and management styles identified in the literature are generally based on two opposing characteristics: the Autocrat and the Democrat.

The autocratic styling is often considered the classical approach to management. It is one in which the manager retains as much power and decision-making authority as possible and often excludes employee participation. It is the most effective style to adopt if environmental risk can only be reduced through detailed orders, as may be the case in an emergency situation, or when recruiting new staff to a job with potential for high environmental risk. However, this style of management may hinder the development of a corporate environmental plan, as it potentially excludes employees from having an input into the continuous improvement of business practices.

The democratic approach encourages staff to be a part of the decision-making process. The democratic manager keeps employees informed and shares problem-solving and project responsibilities. This style requires the manager to gather information and opinion from staff before making the final decision. Such a management style can produce high quality work for long periods of time and therefore is ideally suited to implementing an EMS, which is a long-term process. Many staff members enjoy the trust they receive and respond with cooperation and high morale, all of which is imperative to the continued improvement of environmental performance. However, as with the other styles, the democratic style is not always appropriate. For example, it is often more cost-effective for the manager alone to make a decision and it is an inappropriate style when employee safety is of concern.

There is no one recommended style as each has its own perceived strengths and weaknesses. Each manager is an individual made up of a mix of practical and theoretical learning experiences and activities that occur within an organization and the level of success realized is strongly influenced by the style adopted.

The ideal situation is that at the beginning of each financial year all directors and managers agree and accept the corporate objectives and the associated budgets and their reward incentives for the achievement of the set objectives. This principle of overall acceptance ensures for the most part that all directors and managers have a common goal and are all moving in the same direction. The reality is, however, that individuals perceive objectives differently, as well as the best method of achievement. There is a great deal of interpretation by managers and directors and therefore new environmental strategies and systems are prioritized against other quality safety and operational requirements leading to at best, confusion and at worst, obstruction.

Managers and directors may have accounting, marketing or engineering backgrounds and therefore respond to change in either a proactive or reactive manner. Management style is a complex concept and it is very easy to understand why it can act as a key barrier to the introduction of an EMS, particularly if used as a convenient mechanism for those corporate executives seeking a quiet, conservative life requiring little change (Wheeler, 1993). Even in environmentally committed organizations, managers are less supportive when managing environmental activities than other business activities (Ramus, 2002) and some may even take the view that if a problem is ignored long enough it resolves itself (Kirkland and Thompson, 1999).

A more complex view is that the traditional management concept of viewing the organization as a collection of entities, or strategic business units (SBUs), competes with the growing impression that environmental management requires a holistic view of an organization (Taylor, 1992). Studies by Miller and Freisen (1980), Agyris (1993) and Pfeffer (1996) suggest that managers do not follow rational environmental practice because it falls outwith their 'focus of attention'. Managers in industries more exposed to environmental legislation and policies must first translate these legislative and policy requirements into achievable objectives before implementing cost-effective operational strategies (Currie, 1993).

Key managerial behaviour for the implementation of improved environmental performance

Ramus (2002) outlined six managerial behaviours that improved environmental performance, and ranked them according to their strongest impact. The findings are listed in Table 3.2.

Organizational structure

Organizational structure is a framework that allows corporate strategy to be pursued. The relationship between strategy and structure is important if an organization is to successfully achieve its corporate objectives (Miles and Snow, 1978). An organization does, however, adopt a structure that best meets the demands of a unique set of internal and external pressures.

TABLE 3.2 Key managerial behaviours

Managerial Behaviour	Comments
Environmental communication	• Democratic management style shows that managers are willing to listen to employees' environmental ideas, which encourages creativity and participation. • A non-hierarchical approach encourages employees to communicate directly with other business areas to solve problems.
Environmental competence building	• Provides environmental problem-solving tools. • Supports employee eco-innovation. • Managers trained in environmental sustainability so that they are able to support employees and ensure that they understand the benefit in giving employees time and resources to commit to environmental improvements.
Environmental rewards and recognition	• Environmental awards, bonus pay and regular feedback and praise all encourage employees to improve environmental performance.
Management of environmental goals and responsibilities	• Delegating responsibility to individuals and setting goals and targets directly correlate to improvements in employee environmental performance rates.
Environmental innovation	• Managers who are open to new ideas and have an innovative and experimental approach to problem-solving encourage others to innovate.
Dissemination	• Employee perception of a company's commitment to a written environmental policy has important impacts on environmental actions. • Continuous dissemination leads to continuous motivation towards environmental performance. • Consider the least important because managers are generally poor at disseminating information.

The cross-disciplinary, cross-functional nature of environmental issues leads Roome (1994) to suggest that organizations need to reform their existing structure. The development of an environmentally successful organization requires problems of environmental inertia to be addressed, which are familiar in organizations striving to move from one set of structures, systems and values to new frameworks (Mintzberg, 1987).

The importance of this point is underlined in Pujari and Wright's (1994) study that also finds that there cannot be an effective EMS without a change of structure or organization. A change in an organization's structure or strategy usually takes the form of planners presenting plans to top management for acceptance and resourcing (Piercy, 1989). Piercy (1989) suggests that barriers are created when the plans for change are accepted and attention is then turned to the issue of implementation. He states that at this juncture top management considerably underestimates the costs and problems of getting the new plans accepted.

Additionally, it can be argued that directors and managers have a choice of whether to 'adapt' or 'resist' organizational change. Those who prefer the process of adaptation would modify existing corporate strategies to better match those changes that are occurring externally (Child, 1977). Those directors and managers inclined to preserve existing strategies would be those that would be reluctant to change to external pressure.

For organizations operating in an environmentally sustainable context, or low carbon economy, this traditional view of organizational change is being challenged. The change that is required of an organization committed to sustainability or low carbon principles is deeper. In a low carbon context organizational change incorporates a top to bottom approach to provide the necessary structure and low carbon vision. It also has a bottom-up approach to encourage participation by all employees (Buono and Kerber, 2010). However, the depth of such change requires engagement with staff psychology and belief system. For an organization committed to low carbon principles, it must provide a change framework and monitor, at the employee level, changes of perception and adjustment of individual values (Smith and Sharicz, 2011).

Strategy complexity

Developing a strategy is essential for an organization seeking to align internal and external stakeholder concerns about environmental protection and resource efficiency and the health and well-being of staff (Leon-Soriano *et al.*, 2010). Modern corporate strategies require the full integration of social and environmental aspects, the vision, culture and business operations with the required organizational change (Paraschiv *et al.*, 2012). However, Floyd and Wooldridge (1992) suggest that the more complex the strategy, the more potential friction may be created between achieving the strategic objectives and strategy implementation. Senior management perceive the unsuccessful implementation of new strategies as being the fault of middle management. They are perceived to be either unsupportive or ill-informed of the new direction. Middle managers consider that new strategies tend to be successful if there is a clear set of strategy objectives and priorities and a shared level of understanding and commitment.

The more complex the EMS, the more organizational forces act against implementation. A study by Rothenberg *et al.* (1992) found that effective environmental strategies were integrated with existing corporate strategies that were consistent with organizational characteristics and operating context. The purpose of an EMS is to make complex environmental issues manageable (Kirkland and Thompson, 1999). Unfortunately, managers and other stakeholders are prone to see EMSs as adding to the existing organizational complexity they have to deal with.

As increasing number of environmental standards emerge to provide corporate guidelines for those organizations in pursuit of such goals as energy efficiency, sustainability, CSR, the greater the likelihood of strategic complexity. The larger the organization, the more systems and subsystems become a consideration when implementing an environmental strategy. Many organizational groups or SBUs may have their own set of goals, time perspectives and decision-making processes. Designing strategies without consideration of these substructures and their respective strengths and weaknesses may also hinder the implementation process (Rothenberg *et al.*, 1992). The recognition of different substructures within an organization may lead to a more effective route to environmental strategic change and lead to the identification of additional product or service development opportunities.

System integration

The main difficulty for managers is that 'one size does not fit all' when it comes to implementing a successful system. Each organization is different in terms of its culture and the way it undertakes its business activities. Any new environmental system must fit with the prevailing political, operational and economic needs of each organization (Rothenberg *et al.*, 1992). In addition, some managers may consider that the issue of system complexity is compounded when environmental, quality, health and safety systems are integrated as one system.

Technology

Technology in this context may be used to describe a) the hardware (machinery) and software and firmware (systems and techniques) used to process and present data in a meaningful form or b) the hardware used as part of the business, such as manufacturing and production. In the case of the first context, it has been argued that because of the complexity of information involved, flows of information tend to be more upward than downward (Lorsch and Allen, 1973). Another barrier may occur as a result of managers selecting and preferring personal contact and informal communication to assist with network development and decision-making (Mintzberg, 1987). A key issue for decision-makers is how to react to information received from a source that is viewed sceptically and not trusted and that uses an unfamiliar language.

The transition to a low carbon economy, for example, and the deployment of environmentally sustainable products and services and energy-efficient technology has a significant impact on organizational change. The change is in the approach to investing and financing the many environmental management technology development funding options: its applicability, reliability and availability, at each stage of the technology development cycle.

In the case of technologies used within an organization, environmentally sound technologies are those that ensure efficient resource use with reduced wastage and emissions. Many environmental issues within an organization may centre on the complexities or inefficiencies of technology choice. While innovations and technology improvements are continually being made, new technologies often require new skills and training and a broader infrastructure support in terms of management and maintenance (Hale, 1995). All these things come at a cost in time and finance, and often organizations may be put off upgrading or modifying equipment because the initial cost of investment cannot be regained in a short enough time period or because of uncertainty on which technology to adopt.

Top management commitment

Earlier EMSs were generally seen as being isolated from the main function of the organization (Shelton, 1994). This was due, in part, to the failure of top management who expected the EMS to adapt to the prevailing business culture and, in part, to environmental managers who expected to be accepted into the prevailing organizational culture (Hunt and Auster, 1990; Buzzelli, 1991). Cohen and Levinthal (1990) referred to this phenomenon as an organization's 'absorptive capacity'.

Absorptive capacity is seen as a function of an organization's prior related knowledge. That is, an organization's capacity for assimilating and applying external information for commercial benefit. In the case of environmental management, the absorptive capacity theory would imply that prior knowledge would be limited and therefore the introduction of an EMS would be difficult due to the lack of top management knowledge that could limit the level of commitment.

The difficulty for senior management is that once an investment decision has been made, they rely on feedback information to determine whether the investment decision was a correct one. Often, any negative feedback received before the investment objective has been realized can have a detrimental effect on continued commitment. Once uncertainty has been created on an investment decision, senior management commitment can wane, and additional resource requirements may no longer be viewed as further investment but as an expense (Cohen and Levinthal, 1990). It is at this point that decision making slows and more analysis is required before further resources are committed. The final outcome may be that investment decisions stall or are abandoned completely (Shelton, 1994).

It is claimed that top managers strongly influence the implementation of environmental management. On green organizational issues three levels of top management commitment have been identified. Ghobadian et al. (1998) categorize environmental commitment as:

- Restrained commitment.
- Speculative commitment.
- Conditional commitment.

Restrained commitment refers to companies that may want to make an environmental statement, but do not perceive any real need to follow up this statement with action. The category of restrained commitment can involve symbolic activities such as 'green-washing' taken by some organizations to demonstrate their environmental commitment, while underlying practices and values remain unchanged. Speculative commitment can be applied to those organizations that become leaders in an environmental field because they identify business opportunities such as increased market share, increased profitability, or reduced cost structure leading to competitive advantage. Therefore, speculative commitment can be viewed as opportunistic. Having conditional commitment, organizations can take different actions in different circumstances or countries. That is, an organization's environmental commitment depends on the prevailing business and economic conditions (Ki Loon and Ball 2003). Therefore, a proactive approach may be adopted where such a response is required, for example, investing in environmental technology and pollution reduction systems. In contrast, a more reactive action may be taken when situations allow or the organization is best served.

Barriers in small to medium-sized enterprises (SMEs)

In the case of SMEs the barriers highlighted in this chapter take on a different emphasis due to the particular structure of the smaller enterprise. In a smaller business, the investment of resources needed to implement an EMS is more significant when compared to the overall turnover of the business, but potential improvements are also more significant to the future success of the business.

Hillary (2004) undertook a study of SMEs to determine the benefits and drawbacks to SMEs of implementing an EMS. Three main benefits arising from Hillary's study were identified as follows:

- Commercial benefits: attraction of new business and customers and the satisfaction of customer requirements.
- Environmental benefits: assured legal compliance, energy and material efficiency and reductions in energy consumption and waste production.
- Communication: enhanced image and better dialogue with stakeholders.

However, while these benefits are significant to any business, many SMEs find the barriers to implementing an EMS so great that these benefits cannot be realized. Some of the key SME business barriers are listed in Table 3.3.

The barriers already discussed in this chapter can reduce the impact of EMS or make it unsustainable, but rarely would any individual barrier stop EMS implementation from being initiated within larger organizations. Some of the barriers identified by Hillary (2004) such as lack of assistance, underestimating finance and other resource requirements, additional demands on existing staff to the detriment of the business, often stop EMS implementation at the first stage, so that no further attempt to implement EMS is ever made.

TABLE 3.3 Barriers for SMEs

Resources	• Few SMEs can afford to employ full-time environmental managers, and as a result the demands of EMS implementation take managers away from their main job responsibilities. • Multifunctional role of staff undertaking EMS along with their usual jobs means that little time is available for all responsibilities. • Few general managers have the technical knowledge to implement EMS or the time to improve their knowledge. • High cost of certification and verification which disproportionally penalizes smaller companies. • Underestimation of resource requirements leads to SMEs giving up before they achieve any benefits.
Understanding and perception	• Lack of awareness of benefits may reduce company motivation for implementing EMS. • Lack of understanding may reduce the value of environmental reporting leading to ineffective communication with stakeholders. • Perceived high cost of implementation and maintenance. • Uncertainty about the commercial market value of EMS.
Implementation	• Implementation can be interrupted which lengthens the process and may lead to waning interest in completion. • Inability to see the relevance of all the stages.
Lack of rewards	• SMEs found that components of the EMS failed to meet expectations, such as automatic compliance with regulations, competitive advantage and stakeholder satisfaction.
Attitudes and company culture	• Management instability. • Potential low status of the person leading the EMS implementation. • Resistance to change. • Lack of internal marketing of EMS.
Support and guidance	• Lack of accessible financial assistance. • Lack of consultants with experience in implementing EMS in SMEs. • Inconsistent approach of consultants to EMS implementation. • Lack of sector specific exemplars. • Lack of trade association or support network. • Poor quality information of conflicting guidance.
EMS surprises	• SMEs were often underestimating the resources required to implement an EMS and were unable to rectify the situation with further resources due to the small scale of the company.

Low carbon technology developer funding experiences

Lack of management and communication skills

Besides the general barrier financiers predominantly complain of when funding SMEs is the lack of management skills within a young business. However, within funding institutions there is often a lack of knowledge and management experience. For example, managers or owners in small energy businesses often remark on the lack of dedicated in-house energy management expertise in potential investors or funders to assess and promote low carbon technology projects. This can create uncertainty between financier and business and result in a lack of information on available funding options. In addition, the separation of project responsibilities or decision making, for example different departments for energy expenditures and conservation, can lead to communication gaps and confusion. The coordination and communication between different stakeholders such as financial and technical departments also serves to slow the decision-making process.

Barrier solution

One example of a solution to tackle some of the SME barriers mentioned above is to encourage SMEs' funding applicants to join knowledge networks and provide support with contacts to key stakeholders. Within the Federal Ministry for Transport, Innovation and Technology in Austria there is an umbrella management group which supports the SME applications during the funding application process by creating a network for assistance in the financial evaluation, contract completion and results assessment. Furthermore, the mix of stakeholders involved in the application process work closely together which decreases the communication failure and strengthens the application. SME support networks also play an important role when it comes to co-finance, for example the French Agency for Industrial Innovation (AII) initiates contacts with many business stakeholders, governmental agencies or ministries in many countries.

Lack of capital

SMEs often encounter the barrier of insufficient collateral to secure the requested financial support. In the early stages of technology development it can be difficult for developers to find investors. This puts additional pressure on the technology developer. In some schemes, only projects with sufficient financial resources for project implementation are eligible for funding support. A confirmation from the bank or co-funding institution about its readiness to provide the additional money is almost mandatory in the early stages of technology development.

In addition to simple lack of capital, sometimes adequate capital planning tools are missing to allocate investments properly over the year and be able to react to sudden market opportunities and trends. Many of the financiers find that there are more eligible applicants than financial means within the instrument's budget and therefore even promising projects have to be refused. Most funding programmes have a fixed budget provided by the state and are therefore inflexible to react to more applicants and suddenly changing needs. Moreover, some programmes' small budgets prevent the development of certain technologies which need larger investments, for example larger wind farms.

Barrier solution

The setting of co-financing requirements should take into account the beneficiaries' capacities, such that more lenient requirements on the amount of co-financing can be stipulated in certain cases, and that larger advanced payments can be made available. Some application procedures take a long time, which puts an additional financial burden on the project developer. It is important for SMEs to recover any direct costs arising from participation in planning and consultation processes. This could be overcome by the use of compensatory advance payments made from national public sources to mitigate the acute cash-flow problem they might experience. A revolving fund could be set up to offer preferential terms of interest rates, collateral, service fees and grace periods for loans and bank guarantees needed for co-financing.

Risk assessment

Where renewable energy projects are small, such as solar photovoltaic (PV) and mini-hydropower, the transaction costs are disproportionately high in comparison with conventional power projects. The cost of initial feasibility and due diligence work does not vary much with project size. Therefore, pre-investment costs, such as legal and engineering fees, consultants, and permitting costs have a proportionately higher impact on the transaction costs of renewable energy projects. Furthermore, the generally smaller nature of these projects causes lower gross returns, even though the rate of return may be well within market standards of what is considered an attractive investment.

Generally there is often a lack of technical expertise for risk assessment and technology verification. This, in addition with the high risk in certain projects, is often a reason for setting the requirements high. In the case of a wind project, for example, at least one year of on-site wind speed measurements is usually required before a financier seriously considers any investment. Therefore, funding institutions prefer to engage with experienced construction stakeholders, suppliers with proven equipment, and experienced operators instead of developers with little or no track record.

Barrier solution

With financial instruments certain risks are moved away from project sponsors and lenders to insurers and other parties who are able to underwrite or manage the risk exposure in a manageable way. These financial risk management instruments can help to mitigate the perceived risks and consequently affect the degree and terms of investment into sustainable energy projects. However, constraints on the availability of such risk management instruments come from and depend on the willingness and capacity of insurance and capital markets to respond. Therefore, there are still considerable gaps in providing insurance products for the sustainable technologies on the market. It is the task of the public sector to provide risk minimizing instruments and leverage more private investment.

Lack of awareness

The perception among developers within SMEs is that that they are generally not aware of available technology funding programmes. Additionally, they are aware that they need to close the gap of knowledge on available technologies for potential investors. As mentioned earlier, one of the principal barriers from the technology developer's perspective is the lack of understanding and experience of the technology that exists among commercial lenders. In particular, there is a distinct lack of understanding of technology and its benefits around low carbon technological development.

Barrier solution

Mechanisms to overcome these barriers can include institutional strengthening interventions, such as training and technical assistance to banks, which is offered by IFC (International Finance Corporation, World Bank) before providing guarantees. These third parties are then able to put in place appropriate credit rating and due diligence procedures, or project specific support facilities that share some of the elevated transaction costs. Raising awareness is crucial within smaller commercial lending institutions, particularly in emerging markets, but also in larger institutions. Educating management and board members can help raise the profile of investing in low carbon technology and improve the general understanding of sector trends and market opportunities. To increase access to information in this field one option is to generate investment forums or knowledge networks for different low carbon technologies.

Market failure and funding gaps

Barriers and gaps in funding low carbon technology – the investor perspective

In the LCEGS sector a number of barriers exist that slow or prevent scaled-up investment. Financial structure and scale are a challenge to investment in this sector as well as the usually high up-front capital costs compared with conventional energy technologies. The external financing requirement is therefore high as the level of risk is perceived to be higher. Moreover, the typically small size of technology development projects makes transaction costs disproportionately high (feasibility analysis, due diligence, legal and engineering fees, consultants, etc.) as these costs do not vary significantly with project size. Financiers regard the cost and long-term performance risks of clean energy technologies as being higher than with conventional systems – a perception which often results from a lack of timely and accurate information. Furthermore, the market distortion caused by the pricing of high-carbon fuels, which does not reflect the environmental and social costs they impose, puts most low carbon technologies at a competitive disadvantage and makes them dependent on supportive policy and regulatory frameworks to be financially viable. The specific challenges that contribute to the financing gaps depend on regional, financial, regulatory, and technology differences.

Barriers to the demand side

Technology developers within SMEs often do not want to share ownership and control over their ideas with equity investors and would rather find other financial solutions than accept limits to business growth. Furthermore, being able to evaluate the available funding options and to understand the concerns and needs of investors is essential for entrepreneurs who try to get risk capital funding. Other important factors affecting demand, apart from cost-related issues, are a lack of awareness of the society, and the limited cultural acceptance of new technologies.

Gaps in funding technology innovation

Technology developers, governments, investors and eventually end-users influence the different stages of technology development: research and development, demonstration, pre-commercialization and commercialization. The funding gaps typically begin in the demonstration stage and continue through to the early commercialization phase. The gap is caused by systemic market failures encountered as technologies move through the technology development process. Grant funding, often accessed at the early research and development stage, is rarely available for demonstration activities, particularly where demonstration involves high capital costs. There is even less funding available for the pre-commercialization stage which is also characterized by high-cost activities such as initial and secondary prototype development and testing, site development, supply chain formulation. These trials and prototype retrofits are cost-intensive but necessary to convince investors that the technology is able to perform in real-market conditions.

Research and development is the primary focus of government support in industrial and innovation policy. It is supported by government grants and subsidies. The capital required at this stage is relatively small compared with the level of investment necessary at demonstration and pre-commercialization phases where business-development support and venture capital become more important.

Gaps in funding low carbon technology

SMEs are key players in the progress of low carbon technology markets and in providing services supporting the industry as well as consumers. Most of the clean energy and low carbon technology businesses responsible for technology innovation are SMEs. Their activities range from consultancy services to small-scale manufacturing and assembling, wholesaling, distribution and installation, small-scale sales and servicing. An added value for SMEs is that on the one hand they are often locally based, which can be an advantage when convincing consumers and other organizations to engage with low carbon technology and on the other hand they are flexible, which puts them in a good position to work with larger organizations to support their business activities.

There are only a few government-supported enterprise development and business support programmes in place that are willing to address the critical financing gaps encountered by business start-ups. Investors often consider low carbon technology and SMEs as being high risk and high-cost investments. Therefore, supportive regulatory or fiscal frameworks and financial support is necessary for those enterprises experiencing capital constraints or difficulties in achieving sufficient profit margins in all stages of development, from the concept and early business planning phase through to full operation.

Investors tend to face a steep learning curve with new technologies and as a consequence often stick to conventional, tried and tested technologies. This insufficient knowledge and misconceptions about technical requirements and financial benefits and the coherent lack of trust often result in a high perceived risk, especially when an immediate beneficial impact might not be visible at first glance.

One of the key future objectives of the EU is to become the market leader regarding low carbon technology development and to be seen as the 'global environmental engineer'. Driven by regulations and initiatives such as the Lisbon Agenda and the Environmental Technologies Action Plan (ETAP) the EU is committed to closing the gaps and reducing the barriers to funding low carbon technology by implementing funding programmes as well as providing secondary support such as networking activities. The intention of the EU is that programmes help to create a positive investment climate for technology developers. Examples of these EU programmes are detailed below.

Competitiveness and innovation programme (CIP)

The CIP (2007–2013) has a budget of €3,621 million and a range of actions supporting innovation and SMEs. The development of technology is a cross-cutting theme covering three key areas: Entrepreneurship and Innovation, Information and Communication Technologies, and Intelligent Energy Europe. Within the sub-programme Entrepreneurship and Innovation, €433 million (out of €2,172 million) is earmarked for specific support actions on low carbon technology and is supported through three types of measures:

- Financial instruments in support to investment funds active in technology development with an indicative budget of €228 million (2007–2013).
- Networks of national and regional partners, with an indicative budget of €10 million.
- Pilot and market replication projects, with an indicative budget of €195 million.

One barrier or obstacle is a lack of clear success measures on the performance of invested capital. Many institutions use 'soft' indicators, for example, survival rate of the funded SMEs, energy-saving potential of the developed technology, innovation of products or the number of created work places. Other institutions use 'hard' indicators that are more focused on the financial outcomes of their investment such as return on investment and what additional finance can be leveraged from other funding sources. Performance measurements enable better focusing and allocation of funds and therefore the development and establishment of clear target settings and factors of success. Ultimately, organizations involved, or looking to become involved, in developing low carbon technology would benefit from the diffusion of such knowledge.

Environmental improvement is a political focus and one that is largely incorporated within public sector funding instruments. Low carbon technologies offer strong opportunities for the establishment of new industry sectors with opportunities to create new jobs and grow the national economy. The focus of public funding instruments lies often on classic supportive measures for economic development and growth. Explicit environmental targets which can be achieved through intensified efforts, for example, through development of new technologies and products, are not core targets and also not widespread within the funding institutions.

Availability and accessibility of information on public funding instruments and support institutions is still lacking. The level of detail of publicly available information from these institutions varies significantly and often only very general information is readily accessible. Technology developers therefore face challenges to identify easily the right funding institutions and instruments to apply for. Additionally, divided, overlapping or partially coordinated responsibilities within the national low carbon technology funding network make it unclear which institutions act as funding bodies or advisers and how different schemes can be combined and utilized.

A secondary but interesting aspect is that only very few experts had an overview on the public efforts in funding low carbon technologies in other EU member states or had contacts to similar institutions. A better networking and exchange of knowledge and experience would eliminate many barriers and obstacles in funding low carbon technologies. Possible future steps could be – as environmental problems not only have impacts on single countries but can have worldwide or regional impacts – an intensified networking and collaboration between institutions from different countries. This would result in a better interaction and coordination of targeted actions between European, North American and Asian countries.

Corporate finance

Perspectives on private sector finance

Two key insights among technology developers about private sector financing stand out. Technology developers with in-depth knowledge of the complexities of the financial markets have tended to adopt a more innovative financial model for their enterprise, which improved their access to finance. Also, general market conditions have a big impact on the relative attractiveness of investments in environmental technology development.

Other commonly-cited barriers include: unattractive conditions of investment; the high risk perception of projects is damaging; there is a lack of investor confidence, and developers find it difficult to provide guarantees, collateral or other risk-sharing mechanisms because of the weak nature of their balance sheets.

Larger organizations tend to have more advanced financial models in place than the smaller businesses, and their size and position allows them better access development funds. However, technology developers and SMEs are creating global networks and are building beneficial strategic partnerships with large multinational corporations. For a young enterprise, it is important to take on investors with experience and good connections. These so-called 'added value' traits are in some cases more important strategically than the financing itself.

The developer can provide some guarantee of future success by obtaining a letter of intent from an industrial partner or partners. A private investor or public agency can guarantee the output of the technology to the industrial partner. In this event, the industrial partner faces little risk in allowing the demonstration to go ahead.

Technology developers can increase awareness of their technology using press releases, prizes, competitions, showcases, trade fairs and conferences to boost their public profile.

Developers need a toolkit too, to give them a guide to the financial packages available, and improve skills relating to investor readiness, business management and marketing. A centralized database and matching service would be a great time-saver for technology developers looking for private sector funding providers.

Non-traditional business models could provide an opportunity to unlock some of the wealth of industries and other large organizations but also to take advantage of their skills, experience and networks.

Perspectives on public sector finance

Another key barrier for SMEs or technology developers is the administrative burden and the length of time it takes to go through the support application process. There are so many different funding schemes that technology developers find it difficult to find the instruments that suit their needs the best and are easily dissuaded from future applications due to low success rates.

In North America, for example, the public sector financing bodies are very advanced and offer many complex packages of investment, support, networks and management training, but also verification and validation of the technology, often at a state or provincial level.

Support and information is provided, such as manuals and other aids; feedback on enquiries, opinions and advice; consultancy for preparing the essential documents and giving recommendations to improve the quality of the funding application prior to and during the application process.

Another attempt to tackle some of the barriers mentioned above is by supporting the applicants in joining networks and providing them with contacts to researchers and the industry. The European Technology Platforms have made a strong move towards this, combining researchers, industry, market regulators, and the public sector to determine the future strategic aims and the development needs for specific technologies.

4

EMS MODEL TYPOLOGIES

Summary

EMSs offer organizations a framework with which to implement processes and procedures to reduce the environmental impact of their business activities. The unique mixture of environmental drivers and organizational barriers within an organization requires a customized approach to environmental strategy development. Such an approach involves integrating operational demands with social and environmental demands to meet achievable objectives and targets. Utilizing a number of environmental development and performance evaluation models, this chapter compares and contrasts the characteristics of a number of model options an organization can adopt for strategic development.

Corporate managers are increasingly interested in developing systems and practices for improved environmental performance. Due to growing environmental regulations, government pressures, international certification standards such as the ISO 14000, changing customer demands, and managers recognizing pollution as waste, corporate managers must now develop environmental strategies and policies for their organization and its supply chain partners while being consistent with new regulations (Lansiliuoto and Jarvenpaa, 2008). Consequently, managers are recognizing the importance of systems used to manage environmental practice and performance. However, for managers deciding how to tackle environmental issues, transforming this recognition into the development of an EMS can be challenging.

An organization has many forces driving it to implement an EMS. These forces can arise both from within and outside of the organization and can change during the course of system adoption. Bansal and Roth (2000) identify three categories of organizational change to identify and analyse the forces occurring. The first category incorporates efficiency choices, including reliability, usefulness and updating of an existing production or manufacturing system or other operational improvement change. The second category covers the situation where the adoption of a system has been forced upon an organization from head office. The final category is when the management team within an organization has a desire to try a new management tool. Malmi (1999) found that the forces changed over the course of the diffusion from the efficiency to the fashionable forces.

These organizational forces also include several subfactors such as management style, culture, communications and strategy complexity, emerging as potential organizational barriers to system implementation (see Chapter 3). The internal and external forces within an organization arise and change during its transition through the motivations and adoption of models which affect the implementation of environmental change.

One source of emerging external forces may be regulators and public authorities (Lozano and Vallés, 2007). Additionally, it is suggested (Darnall, 2006) that environmental regulatory and government pressure may be major drivers of managerial environmental action. And that the stronger the regulatory pressure, the greater the requirement for a public authority mandate, as opposed to gentle encouragement to adopt environmental standard certification. Regulatory pressure is considered to be the most influential external factor for organizations in European countries addressing environmental issues (Rintanen, 2005).

However, while regulatory pressure may serve as a driver for environmental change, there are other factors that affect the implementation of an EMS (Rothenberg, 2007). Other forces can originate from stakeholders such as competitors, staff and customers. The customers may, for example, demand the implementation of an ISO 14000 standard for purchase of product or service. Also, entry into international markets, media pressure, product or service mimicry or increased size may drive the development of improved environmental management practices (Bansal, 2005).

Forces other than customers or other stakeholders may also affect environmental management and motivate an organization to obtain an environmental certificate. Pan (2003) found that one stimulus may be a perceived marketing advantage (i.e. image), when many competitors were already ISO 14000 certified, or awareness of the benefits experienced by other certified companies and the avoidance of a potential market export or supply chain barrier.

Three motives for environmental management and ecological responsiveness are identified by Bansal and Roth (2000): the 'competitiveness motive' emphasizes an organization's commitment to improve its long-term profitability; the 'legitimization motive' refers to an organization's desire to improve the acceptance of its actions compared to regulations, norms, values or beliefs; and the 'ecological motive' refers to a motive that stems from the concern an organization has for its social obligations, corporate values and ethics in general.

An effective EMS can help an organization manage, measure, and improve all environmental and social aspects of its operations. It has the potential to lead to greater knowledge and more efficient compliance with mandatory and voluntary environmental requirements. It may also help an organization effect a culture change as environmental management practices are incorporated into its strategic aims and operational activities. One major challenge currently facing an organization is that environmental expertise and data may reside in a parallel information system, not part of a centralized management information system (Sroufe et al. 2000). Environmental systems and projects are ultimately a cross-functional and cross-departmental undertaking which includes the adoption of different operational technologies, practices and priorities.

EMSs offer a framework and accreditation system for organizations to use in creating continued environmental improvements. However, in reality each EMS must adapt to the business environment (Greeno and Robinson, 1992) and so integration of a standardized EMS requires some modification if implementation is to be successful. An organization can have any combination of drivers (see Chapter 1) for implementing an EMS. As drivers for implementing an EMS vary, so the strategies adopted to achieve environmental improvement will differ, the same strategy will not suit every organization. In addition to this, as seen in Chapter 3, the specific barriers arising in an organization determines the route it takes towards EMS and whether full accreditation is ever achieved.

As the ISO 14000 standard series was not introduced until 1996, prior to this date, many organizations had introduced their own environmental procedures, which worked so efficiently that there was no requirement to change to the standardized system. For this reason, standardization is yet to be achieved and there remain a number of different procedures and strategies adopted by organizations all over the world. Such variation suggests that there are many ways in which environmental behaviour can be successfully adopted (Khanna and Anton, 2002).

Types of environmental strategies

An organization can adopt very different environmental management strategies depending on its attitude to the environmental issues affecting it. Some choose compliance and go no further than what is required by legislative and regulatory demands. Others choose to go beyond compliance, to strive for a competitive advantage and ensure that they are recognized leaders of environmental management in their sector.

While the sustainability of different environmental strategies is largely reliant on available resources and employee environmental competencies as discussed in Chapter 3, an organization's environmental culture and its strategic commitment are also key factors. The types of organizational reactions, responding to the introduction of an environmental strategy, have been classified by Azzone et al. (1997) and are listed below:

- Passive, lobbying-based environmental strategy: organizations with a compliance-based attitude which try to influence governments, regulators and customers to delay the implementation of new regulations on the development of new markets. These types of organization react to environmental improvements as threats rather than an opportunity.

- Reactive environmental strategy: reacting to external pressure from 'green movements', governments and regulators, or other organizations outside the sector whose initiatives could be successfully transferred.
- Anticipatory 'green' strategy: carefully considered timing of environmental initiatives providing competitive advantage for the future. Strategies include the early development of technology for the long-term saving of resources.
- Innovation-based 'green' strategy: where the environment is seen as the most important competitive priority and innovation is used as a solution to improve environmental performance and fulfil new market needs with environmentally friendly products.

The anticipatory and innovation-based strategies are those that belong to the most 'proactive' organizations and proactive organizations are those that have the greatest opportunity to benefit from EMS in many ways including achieving competitive advantage (Morrow and Rondinelli, 2002). Ramus (2002) has identified key environmental policies that exist in environmentally proactive organizations (see Table 4.1).

Environmental management using comparative models

In an attempt to improve understanding of environmental management, academics and practitioners have sought to classify corporate environmental behaviour and evaluate performance (Lansiluotto and Jarvenpaa, 2008; Kolk and Mauser, 2002). To this end many models have been constructed in an attempt to identify patterns

TABLE 4.1 Environmental policies which exist in environmentally proactive organizations

Environmental Policy Themes
1   A written environmental policy including specific targets for the improvement of environmental performance
2   Publication of an environmental report
3   An EMS
4   'Green' purchasing policy, including the reduction of unsustainable products
5   Environmental training for all employees
6   Employee responsibility for environmental performance
7   Life cycle analysis policy
8   Management understanding of SD
9   Fossil fuel reduction policy
10   Toxic chemical use reduction policy
11   Using the same environmental standards abroad as at home

for improved organizational 'greenness', environmental performance or competitive advantage. Environmental management models emerged in 1987 (Petulla, 1987). The models were based on traditional classification and categorization techniques to understand social and organizational phenomena. The environmental management models discussed later in this chapter demonstrate a wide variety of organizational and operational characteristics. These characteristics are captured and represented in phase or stage models that describe a development in time of integrating environmental issues into daily business activity. Despite concerns over inflexibility, environmental models have been considered useful in understanding organizational structures and strategies in an environmental context. Since 1987 over 50 environmental management models have been created (Kolk and Mauser, 2002). Models generally fall into two main categories, typology and continuum. Typology models are categorized on an organization's current situation, while continuum models are categorized according to a progression towards environmental performance excellence.

Typology of environmental models

A typology consists of a conceptually derived interrelated set of ideal types. Some of the identified environmental models attempt to categorize an organization's position in relation to its current environmental situation. Unlike continuum models there is no set criteria used to categorize organizations into mutually exclusive groups. Taken as a snapshot, an organization is classified by the way it addresses environmental issues at that moment in time. There is no attempt to identify future environmental development or response options. Examples of some of these models are listed in Table 4.2 with a brief overview of the categories used.

TABLE 4.2 EMS model typologies

Name	Model Title	Categories
Steger (2000)	Environmental strategies	Indifferent, defensive, offensive, innovative
Lee and Green (1994)	Strategic options for green product development	Do nothing, generic strategies, diversification, remedy, tonic, bread and butter, nimble, leadership, pioneer
UNEP (1995)	Environmental strategy transitions	Complacent non-innovator, complacent innovator, responsible non-innovator, responsible innovator
Rondinelli and Vastag (1996)	Classification of environmental policies	Reactive, proactive, crisis preventive, strategic
Winn and Angell (2000)	Corporate greening	Deliberate reactive greening, unrealized greening, emergent active greening, deliberate proactive greening

Continuum models

A continuum is a linear classification scheme that identifies a continuous, non-discrete, structure in time. Continuum models are classification schemes that characterize phenomena into mutually exclusive groups. Each organization can be identified using set criteria, which positions it within a particular stage or phase on a developmental scale, or continuum. Some examples of continuum models are given in Table 4.3, although not all are discussed here. One example of a continuum model is that of Hunt and Auster's (1990) five stage model.

Environmental stage and performance models

In order to provide a broad understanding of the different types of models and their characteristics, the following five models have been used for discussion and comparison of application:

- Typology: greening model (Winn and Angell, 2000).
- Typology: classification of environmental policies (Vastag *et al.*, 1996).
- Continuum: environmental contexts (Azzone and Bertele, 1994).
- Continuum: strategic options model (Roome, 1992).
- Continuum: five stages model (Hunt and Auster, 1990).

TABLE 4.3 Examples of continuum models

Name	Model Title	Categories / Stages
Hunt and Auster (1990)	Stages of Environmental Management	Beginner, firefighter, concerned citizen, pragmatist, proactivist
Greeno (1991)	Posture towards environmental issues	Problem-solving, managing for compliance, managing for assurance
Müller and Koechlin (1992)	Stages of Environmental Strategy	Ostriches, chicken lickens, green hornets, robin hood
Roome (1992)	Strategic option responses to environmental pressures	Non-compliance, compliance, compliance plus, commercial and environmental excellence, leading edge
Azzone and Bertele (1994)	Environmental contexts	Stable, reactive, anticipatory, proactive, creative
Dodge and Welford (1995)	ROAST Scale	Resistance, observe and comply, accommodate, seize and pre-empt, transcend
Hart (1997)	Environmental strategy	Pollution prevention, product stewardship, clean technology

Corporate greening model

Winn and Angell (2000) constructed a 'corporate greening' typology model from mail surveys and interviews with 135 managing directors from German manufacturing organizations. The study describes the attitude of organizations towards 'corporate greening' and categorizes them into four types.

Deliberate reactive greening

Organizations that undergo 'deliberate reactive' greening engage in specific environmental activities only when forced to do so by regulatory authorities. They have a weak top management commitment to the environment, as the environment is not seen as the organization's responsibility. Environmental considerations are not seen as part of functional or operational decision making. No monitoring activities are performed and performance is never measured so that new regulation or emerging environmental issues may often come as a surprise to the organization.

Unrealized greening

Organizations that undergo 'unrealized' greening have a lack of formal planning and monitoring, leaving the organization often unable to cope with new developments in legislation. These organizations may have environmental policies that consider the environment in decision-making and top management commitment, but they do not have a proactive approach to implementation of an EMS and do not include environmental considerations in their organizational goals. There is also very little environmental innovation undertaken and product design is not executed with the environment in mind.

Emergent active greening

Organizations that undergo 'emergent active' greening perceive opportunities from environmental activities such as cost savings and have a proactive approach to environmental activity up to a middle management level. These organizations take responsibility for the environment through 'green' product design and the organization monitors its own environmental performance and engages in planning and external monitoring activities with prevention of environmental issues in mind. The organization generates environmental product innovations but the environment is not systematically considered in all decisions across all functions and no management commitment is shown above middle management level.

Deliberate proactive greening

Organizations that follow a 'deliberate proactive' greening have a systematic approach to environmental activities. They put environmental commitment and implementation high on their agenda and are environmental innovators. The environment is a consideration in all functional decisions, and material flows analysis is a tool that is used to design product with the environment in mind. Top management is committed to the environment and to SD and these organizations are capable of preventing environmental issues by using systems to plan, monitor and anticipate.

Classification of environmental policies

In order to clarify the environmental management approaches taken by organizations, Vastag *et al*. (1996) group organizations into differing typologies. These classifications consist of four identified groups labelled A to D (Reactive, Proactive, Strategic and Crisis Prevention). Each group is detailed as follows:

- Group A, reactive environmental management: this group refers to organizations that operate in industry sectors that have low environmental emissions, which do not provide a high environmental impact risk and are unlikely to affect a great number of people. This group also contains organizations that use non-exhaustible resources and little energy, and limited transportation. Typically, industry sectors that use well-developed technologies, such as textiles and food producers may successfully exist within this group as environmental management simply calls for regulatory compliance rather than the development of environmental contingency plans in the event of an accident. As a consequence, the responsibility for environmental compliance can safely remain within the existing middle management capability.
- Group B, proactive environmental management: this group refers to those organizations in industry sectors where the use or development of technologies create high levels of environmentally harmful pollution. However, with responsible location and sound environmental infrastructure the environmental impact of business activities can be significantly reduced to acceptable minimum levels. In addition, in order to reduce the potential risk of environmental damage, managers need to anticipate and respond to changes in legislation, technology and customer opinion. In this group, environmental management tends to be decentralized and concentrated at the source of the environmental risk, that is, manufacturing plants.
- Group C, strategic environmental management: this group consists of organizations in potentially high polluting industry sectors that operate in a context that increases risk through changes to external conditions and particularly change in media and general public attitude. For example, chemical manufacturing or waste incinerating organizations based within city boundaries or highly populated areas are included in this group. Under these circumstances, environmental management and responsibility should be implemented at senior management level and within corporate responsibility policy. These organizations are often involved in initiatives that go beyond compliance and

have an aggressive approach to preventing environmental damage. Environmental strategies must be well defined, well communicated, well monitored and transparent to internal and external stakeholders.

- Group D, crisis preventive environmental management: this group of organizations does not contain great polluters due to the small volumes of resources or because pollution is indirectly made. Examples of these industries include the tourism industry, electric energy plants, and fast food chains. However, any pollution that is directly created will affect large numbers of people. A crisis prevention approach is therefore best suited to business operations as the occurrence of an environmental impact is likely to be small but far-reaching in terms of scale. Environmental strategies include a mixture of improving technological procedures to minimize risk and educational campaigns to ensure that the public does not perceive an exaggerated danger in the event of any pollution.

Environmental contexts

In responding to environmental issues Azzone and Bertele (1994) suggest that an organization may adopt one of five environmental contexts:

- Stable.
- Reactive.
- Anticipative.
- Proactive.
- Creative.

A 'stable' context results in management having a lack of awareness as to the organization's impact upon the environment. Organizations in a 'reactive' context merely respond to environmental issues and environmental management evolves slowly. An 'anticipative' context occurs through strong public awareness and pressure that speeds environmental change to reduce corporate risk. The 'proactive' context is characterized by an organization's drive for environmental change through available technologies, while a 'creative' context is the search for technology that aids environmental improvement or energy efficiency.

- Stable: environmental legislation has extremely limited consequences for an organization due to the nature of the business so that the frequency of introducing a new EMS or standard is very low. This is more often the result of a lack of perception of environmental issues within an organization coupled with a lack of obvious environmental risk due to the nature of the business activities and lack of customer demand for improved environmental performance.
- Reactive: attention to environmental issues is limited. Organizations that work with hazardous operating conditions comply with legislative and regulatory demands. The response to change is slow and is only implemented with changes

in legislation. Any change to environmental legislation is seen as intrusive and is implemented slowly, as and when required and not to disrupt operational activities.

- Anticipative: this response includes those organizations that have large numbers of more environmentally aware customers and therefore need to meet environmental standards demanded of the customer. Environmental issues become the source of technological innovation usually as a way of meeting more demanding industry standards particularly for organizations operating in Europe. Therefore, innovation is focused predominantly on energy efficiency and process improvement rather than product enhancement.

- Proactive: green consumerism, the green pound and economic sustainability are expressions of a knowledgeable consumer who has an influence on manufacturing process improvement and innovative product improvement coupled with reduced environmental impact. This includes smaller niche markets, such as the detergent industry, but is growing due to the continuous increase in green consumerism.

- Creative: this is best characterized by sectors such as the plastics industry where the general public is aware of the environmental problems associated with the industry, but the best ways forward cannot be agreed upon. Technology is generally agreed to be the best solution, but even the best solutions have their drawbacks, (e.g. with the plastics industry, recycling is often not energy efficient and biodegradable plastics produce toxic by-products).

Strategic options model

The strategic options model proposed by Roome (1992) suggests that an organization has five broad theoretical options for its environmental strategy: non-compliance, compliance, compliance plus, commercial and environmental excellence and leading edge. These five strategic classifications are based on how an organization can react to environmental legislation, external and internal organizational pressures:

- Non-compliance: this strategy is one of default whether through competing objectives, lack of managerial vision or cost constraints on an organization imposed by head office. The strategic approach is taken either consciously or unconsciously by senior managers; either way it is the preferred strategy and a choice made by management not to incorporate environmental imperatives unless absolutely necessary.

- Compliance: an organization that reacts to the minimum of environmental legislative and regulatory requirements adopts a compliance strategy. As legislation lags behind environmental issues, so an organization adopting this strategy could be considered to be backward thinking and unaware of future environmental developments that may provide a competitive advantage through new product development or waste and energy efficiency opportunities.

- Compliance plus: an organization that takes a proactive position on environmental management is classified by Roome (1994) as having a compliance plus strategy. Developing such a strategy requires that EMSs are integrated into an organization's business operations and strategy framework. The key difference between a compliance and compliance plus strategy is for management to move from a reactive to a proactive management style. Such a movement would require the involvement of senior management to challenge existing management strategies and promote organizational change through environmental improvement.
- Excellence: strategies four and five are captured here. The fourth strategy points to commercial and environmental excellence where organizations use the best business, quality and management strategies to demonstrate good management practice. The fifth strategy is classified as leading edge: here an organization using this strategy signals a specific, or specialized, form of environmental practice such as SD or CSR. An organization adopting this strategy is applying leading edge thinking in environmental management and generally sets the standards for other organizations to follow.

The importance of Room's (1994) study is that he identifies the need for organizational change to ensure the acceptance of an EMS. He recognizes the need for planned programmed change to move compliant organizations to compliance plus or excellence organizations. He argues that compliance strategies are based on clean technology techniques such as waste minimization and energy efficiency and are built around the requirements of environmental legislation or regulation.

The five-stage model

One of the earlier continuum studies on the development stages of environmental management was undertaken by Hunt and Auster (1990). They investigated how organizations managed pollution control and reduced exposure to environmental risk. During the course of this study it was found that organizations were spread along an environmental developmental continuum. At the lower end of the continuum an organization was found to have either no environmental management strategy or a strategy that was so constrained by the lack of resources or operational dictates that it was rendered, largely, ineffectual.

The 'five-stage' model suggests that each stage of environmental development represents a generic characteristic, and the specific requirements of each stage vary according to the type of business, the range of potential environmental problems, the size of the organization and the complexity of the corporate structure. Of all the models identified, Hunt and Auster's (1990) model would provide a strong basis from which to determine the stage of an organization's environmental development.

The models discussed in greater detail in this chapter focus, for the most part, on the external factors that affect organizational change and identify the external pressures that would have an influence on the type of EMS being introduced or adopted. Hunt and Auster's (1990) model, in contrast, explores the (five) stages of organizational internal environmental development and these are detailed as follows.

Stage 1 – The beginner

At the lower end of the continuum organizations without environmental management strategies were classified as beginners. Organizations with strategies based upon extending the responsibilities of a senior engineer or operational manager to include environmental management were also classed as beginners. Those organizations involved in banking and insurance services and which view themselves as having little or no impact upon the environment are also classified as beginners.

According to Hunt and Auster's classification, the criteria for a beginner were used to describe a situation where senior managers within an organization consider environmental management to be unnecessary, and generally top management support and resource commitment is perceived as minimal at best. Therefore organizations at Stage 1 of environmental development are seen to offer no solutions or commitment to reduce their exposure to environmental risk.

Stage 2 – The firefighter

The second stage of the five-stage model identifies the firefighter organization. At this stage of environmental development operational activities are focused on firefighting. This is a reactive approach by management whose strategy is one of resolving environmental issues as and when they occur. Generally, organizations with this kind of strategy rely on a small environmental team to address environmental issues as they occur and prioritize on the basis of immediate risk. Such a strategy tends to emerge from the allocation of minimal resources and, due to the necessity to prioritize environmental issues, an organization at this stage is constantly open to serious environmental risk from undetected issues that have yet to surface.

Stage 3 – The concerned citizen

Those organizations aware that environmental issues are part of business activities take a compliance stance and use environmental specialists to identify and address potential areas of environmental risk. An organization adopting the strategy of the concerned citizen is aware of its environmental regulatory and legislative responsibilities and ensures that sufficient resources are made available to monitor and report on the likelihood of environmental risk.

Stage 4 – The pragmatist

At the pragmatic stage organizations are no longer reacting to environmental issues but are taking time to plan and manage the business activities that reduce the likelihood of an issue arising. The environmental departments of a pragmatic organization are well equipped with skilled personnel, have sufficient resources and managers have the authority to ensure effective response.

An organization achieving stage four was considered to be very organized and efficient in reducing environmental risk. There were good recording and monitoring and auditing systems in place to support the operation of an EMS. Despite these achievements it was considered that even at stage four, environmental management was still not top priority. There was still a risk at this stage that funding for environmental development may be tenuous and the impact of the environmental team over operational necessity was limited. Organizations with such a pragmatic approach to environmental management were those that were in industry sectors such as the petrochemical and manufacturing sectors and attracted intense scrutiny from regulatory bodies and environmental action groups.

Stage 5 – The proactivist

In a proactive organization the environmental department is staffed with skilled, knowledgeable and motivated individuals who take the concept of environmental management beyond compliance, policing and prevention. Employee training and awareness programmes extend across all management levels and are an integral part of daily activities of departments and business units. There are clear objectives, targets and systems that facilitate achieving environmental goals built into each operational area. There is also strong support for environmental management from senior managers and this support is communicated through an organization's direct reporting channels, formal and informal staff meetings and discussions.

Hunt and Auster's (1990) five-stage model is similar to Roome's (1992) strategic options model where Roome also classifies organizations into those that do not comply, those that comply and those that exceed their environmental responsibilities.

Environmental contexts

An integrated approach

Traditionally, environmental management strategies within organizations have been compliance oriented to comply with environmental legislative requirements, organizations established small environmental units to deal with legislative and regulatory requirements. As a consequence, these environmental units failed to be sufficiently effective. The environmental units created were found not have an operational or staff function; this meant they had little impact upon their own business units. These units were perceived by management as having no relevance to daily operational activities and therefore of low priority when allocating scarce resources to implement environmental procedures.

While it could be suggested that operational managers were guilty of staying within their knowledge comfort zone, environmental managers were also guilty of sitting back and waiting for recognition and acceptance by operational managers instead of taking the time to explain and demonstrate the benefits of an environmental policy to operational activities. Some managers may have felt that their cause was right but failed to consider that the organization is in business to be financially sustainable and make profits for its investors. It is a business culture that supports essential profit-making activities above corporate good causes.

The creation of such tensions was perceived as the 'Green Wall' (Shelton, 1994) which some organizations' environmental strategies were coming up against and stalling. The green wall grew out of persistent differences in language and culture (Haveman and Dorfman, 1999). For the most part, management recognized that their environmental management strategy was not progressing smoothly but rather than addressing the reasons why, available resources were either redirected or the initiative was abandoned completely. In short, the creation of the green wall was the result of an uneasy fit between strategic environmental management, the business functions and the inability of environmental champions to sell the environmental benefits to other corporate managers.

Some of the symptoms of hitting the green wall were identified by Shelton (1994) and detailed as follows:

- Corporate downsizing.
- Tight financial controls.
- Competing management strategies, i.e. TQM.
- Poor communications.
- Clash of business cultures.
- Poor planning of environmental initiatives.
- Unrealistic environmental objectives leading to loss of credibility.

All of these factors, to a greater or lesser extent, act against the successful implementation of an environmental strategy. According to Shelton (1994), an organization's environmental management strategy has to follow one of two paths: either a path towards an interactive and integrated environmental management strategy or a path towards the green wall (Figure 4.1).

A successful integrated system depends on building a stronger understanding of the relationships between business and environmental objectives (Haveman and Dorfman, 1999). An environmental management strategy that fits with the existing business strategy and uses business language instead of 'environmentaleze' and was an integral part of daily operational activities would be accepted by an organization. A stand-alone environment unit, using environmental language,

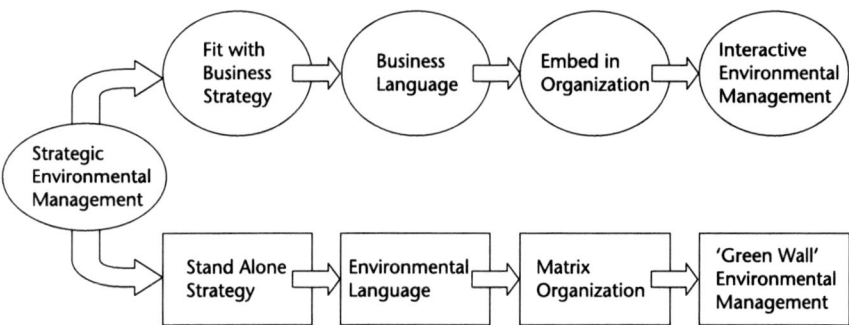

FIGURE 4.1 Shelton's two paths to environmental management

operating within a matrix organization would inevitably come up against the green wall, and stall.

Operational transformation

Research suggests that there are actually two levels to business integration. The first level involves employee awareness and accountability on environmental issues, which is covered by ISO 14001 target setting and auditing and reporting systems. The second level involves the integration of environmental considerations into core business systems and processes (Haveman and Dorfman, 1999). IS0 14001 certification requires organizations to integrate strategic environmental management into business activities through the elements of an effective EMS (Cochin, 1998). Importantly, however, the ISO 14001 standard is a process, not performance, standard so that it does not mandate an organization's optimum environmental performance level but describes a system to aid the organization in achieving its own environmental objectives (Melnyk *et al.*, 2003; Ammenberg and Sudin, 2005). This gives an organization the option to decide to what level it wishes to pursue environmental performance, giving it flexibility to develop at its own pace. However, it also means that obtaining environmental certification is no guarantee that deeper operational transformation will occur, or that a cultural change towards an ecological position will be achieved (Ramus, 2002). An organization often has a narrow perception or knowledge of its own environmental impact, which can often limit EMS to site-specific activities. Operational practices, those operations that an organization is based on, are central to environmental transformation (Ammenberg and Sundin, 2005). Examples of operational practices are:

- Emission filters and end-of-pipe controls.
- Process and production design (Design for Environment), planning and control for reduced energy and natural resources consumption.
- Acquisition of clean technology.

- Preference for green products in purchasing.
- Environmental criteria in supplier selection.
- Shipment consolidation and cleaner transportation methods.
- Recyclable or reusable packaging.
- Recuperation and recycling systems.
- Responsible disposal of waste and residues.

(González-Benito and González-Benito, 2005)

Focusing on operational transformation requires changes in all production and management activities within an organization to enhance its environmental performance. Experiences from countries such as Austria, the Netherlands, Sweden and the UK show that manufacturing environmentally friendly products or services can reduce manufacturing costs (Hui *et al.*, 2001). However, these changes may not be perceived by customers or regulators, and cannot therefore be based on commercial interests (González-Benito and González-Benito, 2005).

Design for the environment

Design for the Environment (DFE) is a design competency that is currently emerging and a key to sustainable innovation and operational transformation. DFE is the adding of environmental attributes or quality factors to a product and service throughout its life cycle. Examples of ways in which DFE has been used are in facilitating automation, eliminating packaging, biodegradable or easily removed labelling to enable easier recycling of existing products (Eagan and Streckewald, 1997).

Product-oriented EMSs (POEMS)

While EMS is a process by which many organizations may first be introduced to the need for environmentally transforming operational processes and DFE, researchers have taken the EMS further to ensure that product design and development is fully integrated into the EMS. POEM is an EMS with a particular focus on the continued improvement of a product, through the systematic integration of eco-design into the organization's strategies. The POEM system started from a policy introduced by the Dutch government with the intention of changing the behaviour of producers. A POEMS system may be based on an existing EMS but adds a much stronger emphasis on life cycle analysis and product design. The effectiveness of a POEMS model along with its integration into an EMS is not yet fully understood and requires further study. The barrier to implementing POEMS is often the limited availability of life cycle data or the lack of knowledge within an organization to use life cycle data. However, it is clear that POEMS may be used to complement existing EMS to aid a better understanding of material flow analysis and therefore promote greater operational transformation. The POEMS model is generally based on the same Deming PDCA (Plan – Do – Check – Act) management cycle as EMS

(Ammenberg and Sundin, 2005). The main stages of a POEMS model are outlined in Table 4.4.

Hutchinson (1996) developed a checklist of criteria that demonstrates the integration of environmental policy with business strategy and which is outlined below:

- A statement from the organization's board showing commitment to integrating environmental management with business strategy and explaining the measures required to achieve this.
- Priority is given to the health and safety of employees, customers and the wider community, rather than to profit.
- The design of products, processes and services is influenced by environmental policy in an explicit way.
- Green and socially responsible purchasing is pursued to avoid using scarce resources, harming endangered species or supporting oppressive regimes.
- Waste recovery through an operating policy of minimizing waste or reuse and recycling of unavoidable waste, with a long-term aim to produce zero waste.
- Reduce pollution at source and monitor emissions with improvement targets set to reduce all pollution.
- Reduce the usage of toxic substances and give special care to the handling and disposal of toxic substances that cannot be avoided.
- Packaging and products (where appropriate) are recovered and reused and recycled.
- Environmental savings are made integral to the business' accounting and budgeting procedures.

TABLE 4.4 Main steps in a POEMS model

Stage	Name	Key Outcomes
1	Product-specific environmental review	- Identify key aspects and impacts. - Review of DFE capabilities. - Review of product development. - Market investigation.
2	Responsibilities and procedures	- Definition of roles, responsibilities and authorities for product development. - Establishment of policies, objectives and targets. - Revision of product development processes. - Establishment of procedures for staff involved in product development and other product related activities.
3	DFE projects	- Development of environmentally compatible products with competitive price, performance and quality standards.
4	Audit/evaluation	- Revision of existing procedures and products aiming for continual improvement.

- Environmental training is provided when appropriate with manager rewards for environmental improvements integrated into the organization's existing appraisal system.
- Continued environmental performance is considered equal to continued business improvement.
- Costs for environmental damage are included in the organization's existing financial reports.

In 1999, Haveman and Dorfman looked into the integration of business and environment in the organization SC Johnson and came up with a list of just three themes that summarizes the features listed by Hutchinson (1996).

- Redefining environmental management issues as materials use issues: organizations learn to see releases and other environmental management issues as materials use issues.
- Aligning environmental issues with key business goals: organizations rethinking and redesigning business goals to accommodate environmental goals.
- Designing consistency into the management system: a key issue is ensuring that environmental improvement is compatible with other management goals. The management system should not send conflicting messages; it should reconcile economic and environmental objectives.

Haveman and Dorfman (1999) also acknowledge, however, that integrating environmental issues into the core business requires environmental trade-offs because not all environmental benefits can be achieved without economic cost. Examples of such trade-offs are listed below:

- Reduction in packaging leads to less protection.
- Changes in packaging increase the opportunities for shoplifting the product.
- Environmental modifications slow down manufacturing.
- Environmental modifications require additional training of staff.
- Environmental beneficial materials increase the cost of production.
- Environmental modifications limit the ways in which equipment can be used.

Organizational reform

The cross-disciplinary, cross-functional nature of environmental problems creates a drive for organizational reform and revised management thinking (Roome, 1994). And, according to Peters and Waterman (1982), establishing individual and shared responsibility for corporate environmentalism should be replaced by more flexible organization and a matrix of formal and informal networks. An environmental policy being introduced into an organization with a strong formalized planning process needs to overcome the inertia of an organization moving from one set of structures, systems and values to establish new ones (Mintzberg, 1987; Quinn, 1980).

Traditionally, structural change within organizations requires some form of organizational adaptation (Mintzberg, 1987; Ansoff, 1991). This traditional management thinking prefers incremental change when introducing change into an organization (Quinn, 1978). Senior management consensus is that incremental change, if done properly, can overcome the resistance to change. In some organizations, however, incremental change can be too slow to overcome a particular problem (Quinn, 1980). This would tend to support Shelton's 1990 environment theory of how EMSs when being implemented can stall or fail completely.

Prior to any environmental considerations most organizations would have established formal strategy and structure for internalizing and meeting regulatory and organizational goals. These existing strategy and structures would have been formed over a number of years by the organizational culture and history (Schein, 1984). The introduction of new EMSs challenges the traditionally conservative culture of most organizations (Smith, 1992). This turbulence pressurizes 'organizations to redefine their goals, to modify organizational boundaries and structures, to implant new value systems and to recognize, and support, specific types of management competence' (Roome, 1994: 66).

Mintzberg (1987: 67) defines strategy as being 'both a plan for the future and a pattern from the past'. He explains that intended behaviour for an organization is based on past patterns of previous action. While he admits that strategy may not have a pattern initially, and new patterns may emerge from that intended by the strategy, he contends that strategy is a craft that requires 'thought and action, control and learning, stability and change' (Mintzberg, 1987: 73). The goal of a manager, in the first instance, when introducing new strategies, is to manage stability not change (Mintzberg, 1987). He adds that when stability is achieved the job of the manager is to know when and how to introduce change.

The suggestion that an organization's past provides important clues for management which are critical for the organization's future success is supported by Greiner (1972). However, while Mintzberg (1987) advocates that management can and should maintain stability while gradually introducing change, Greiner (1972) puts forward the theory that evolution and revolution are the main determinants of strategic change. He adds that these changes are brought about as an organization develops through size of organization, type of market and rate of organizational growth.

To the strategic theories proposed by Mintzberg (1987) and Greiner (1972), Quinn (1978) introduces 'logical incrementalism' to identify and aid the emergence of a new strategy. He argues that it is impossible to predict all the forces and events that shape the future of an organization. And he adds that the best strategies are built on a sound resource base and organization posture that provides rigidity, from which to

> proceed incrementally to handle urgent matters, start longer term sequences whose specific future branches and consequences are perhaps murky, respond to unforeseen events as they occur, build on successes, and brace up or cut losses on failures. They constantly reassess the future, find new congruencies as events unfurl, and blend the organization's skills and resources into new

balances of dominance and risk aversion as various forces intersect to suggest better – but never perfect – alignments. The process is dynamic, with neither a real beginning nor end.

(Quinn, 1978: 18)

The problem with Mintzberg's (1987) theory, according to Boeker (1997), is that as top management structure become more stable senior managers are less likely to deviate from the current course of action, particularly if the organization is enjoying good performance. Existing literature (Reed and Buckley, 1988; Tsai and Child, 1997; Gallarotti, 1995) largely delineates strategy implementation as a series of individual components that need to be considered when introducing organizational strategies. These individual components include strategy–structure fit, resource allocation, executive style, organizational theory, management information systems and budgetary control.

There is little evidence, however, as to how these components react in environmental strategy implementation, or whether one component plays a more significant role at a particular phase of implementation. Patel and Younger's (1978) study explores the differences between the concept of corporate strategy and business unit strategy. They found that corporate strategies were primarily focused on business unit configuration, management systems, and financial transactions with a view to corporate growth and profitability. Business unit strategy, they found, was primarily concerned with competitive advantage.

5

EMS MODELS FOR ORGANIZATIONAL PROFILING

Summary

It is right to state that, generally, the dynamic of a workplace, and the attitude of staff in today's organizations particularly in Europe, Asia and the USA are positively disposed to environmental progress and the pursuit of innovative methods to 'green' the organization. However, there remain many other organizations, in many countries, most particularly in developing countries that are still fighting, pushing and in transition to achieve a commitment to environmental management.

It is argued that, as guides to the degree of effectiveness of EMSs and their development within organizations, environmental models (discussed in Chapter 4) are useful. However, they may also be seen as being two dimensional in scope, meaning that they do not explore the dimension of competitive advantage when integrating an EMS with the existing business strategy. This chapter explores six examples of organizations that have operations on a national and international scale. The cases have been used to facilitate the development of four new models that link environmental management with business strategy in different ways. In an attempt to understand the four EMS models and the profiles of organizations that adopt them, six socially constructed case studies have been used to paint a picture of daily activities within each organization and how these activities affect the pursuit of an EMS. The name of each organization has been omitted. However, the comments and views of staff and managers, their thoughts and actions on how they relate to each of the four models have been included.

An organization's management team commit to EMSs for a variety of reasons. Some do not see the need for an EMS if an existing quality or health and safety system will do. Some recognize the economic and reputational importance of a system that is fully integrated with its business activities. The majority are somewhere between these two approaches.

For each EMS model type detailed in Chapter 4, the following four key organizational characteristics can be used to profile organizations and their approach to adopting an environmental management system:

- Organizational profile.
- Operational advantages.
- Operational disadvantages.
- Organizational barriers.

From the evidence of the following six case studies four categorical models have been constructed to demonstrate the profile of each organization based on how it engages with an EMS. These models categorize those organizations that are 'devoid' of an EMS or have a system that is 'isolated', 'devolved' or 'integrated' within the organization.

Illustrating the classifications of devoid, isolated, devolved and integrated EMS models in this way offers a broad insight to assist managers to understand the implications of each of the different approaches.

Devoid EMS model

The Devoid EMS model, characterized in Table 5.1, is a profile applied to those organizations that have an informal approach to environmental management, and have a system that has not been independently certified. Included in this model classification are those organizations that are currently in the process of implementing an accredited EMS but have yet to complete the process. Many organizations in developing countries such as Mexico and some Asian countries follow international environmental standards such as ISO 14001 but due to a lack of a national certification or accreditation body lack independent system audits or standard verification. Lack of certification means that the EMS is not being monitored and verified objectively, giving an organization little, or no, assurance that its EMS is working sufficiently well to ensure even basic compliance (Morrow and Rondinelli, 2002).

Organizations in the process of implementation do have a lower environmental risk and reduced operational disadvantages. There is a risk, however, that the longer the time taken for installation the more likely the process will stall or fail (Shelton, 1994). In the worst case scenario, all the resources and effort that has gone into implementing an EMS may be wasted if environmental compliance as a minimum cannot be evidenced.

Organizations can operate waste monitoring and energy efficiency schemes without requiring an accredited EMS, thereby accruing some benefits. However, the number of operational disadvantages, listed in Table 5.1, resulting as a consequence of not having an accredited EMS, adds up to a high risk strategy. Many of those organizational barriers identified in the literature and discussed in Chapter 3 are likely to occur in an organization devoid of an EMS.

During an economic downturn, there is a strong probability that an organization with a devoid EMS model is first to delay or abandon the pursuit of an EMS; scarce resources are often redirected to critical operational activities. The implementation of an accredited EMS is renewed when economic conditions improve and as such there is a high risk of an environmental incident occurring.

Management may consider that an operational benefit can accrue by redirecting scarce resources to operational, as opposed to non-operational, activities. Due to the number of pieces of environmental legislation and regulation currently in force (Ball and Bell, 1997) and the lack of operational efficiency resulting from not having an accredited EMS, it is considered that there are few if any operational advantages to be gained from taking this approach. From the number of organizational barriers identified in Chapter 3 the key barriers that can occur in organizations characterized by devoid EMS model are detailed in Table 5.1.

Case study: Organization A

Top management commitment

An organization with a devoid EMS model profile finds that their senior or top managers' commitment to environmental management is minimal; the environmental

TABLE 5.1 Devoid EMS model

EMS model	Devoid.
Organizational profile	No accredited EMS.
	In the process of installing an accredited EMS.
Operational advantages	None.
Operational disadvantages	High risk of heavy fines.
	Adverse publicity.
	Lack of shareholder investment.
	Loss of credibility.
	High investment risk.
	Loss of competitive advantage.
	Loss of market share.
	Not accepted on many major supply chains.
	High risk of environmental incident.
Organizational barriers	Management commitment.
	Communication.
	Available resources.
	Management style.
	Incentives and controls.
	Credible plans.
	Company culture.

manager within such an organization says that 'senior managers are supportive of an environmental management system but only if it is implemented using the minimum of manufacturing resources'. Organizational directors and managers do not consider that their corporate activities pose any serious environmental threat. They believe that an EMS is required for manufacturing activities only. The EMS is therefore supported but only on a minimum cost basis.

Organizational structure

Managers in an organization with a devoid EMS consider that its structure has no impact on the environment and that any environmental impact or risk that may emerge only affects the manufacturing department. In short, too many managers and directors with a devoid EMS view the implementation process as a hindrance. An EMS is often seen as consuming valuable production or manufacturing resources for non-operational activities. Quality and health and safety management systems are viewed as being adequate for monitoring operational activities.

System integration

In a devoid EMS classification, organizations implementing management systems such as ISO 9001 and ISO 14001 are perceived as 'stand alone' systems, there to perform a specific task. These systems are generally seen as part of production or manufacturing activities and the organization does not integrate the EMS with other management systems. The lack of commitment from staff to the EMS is as a direct result of a lack of integration of the EMS. 'Senior managers often send the wrong signals to staff; at present there are two manufacturing staff members who are tasked with the responsibility of implementing ISO 14001, but the environmental manager feels that these staff are tasked with the implementation until other operational commitments call them away, they are then replaced with two different staff members. Consequently, implementation continuity is adversely affected as often replacements are not immediate.'

Managers believe that the existing quality and health and safety systems take care of any environmental legislative requirements. For example, Waste Electrical and Electronic Equipment (WEEE) and Packaging legislative changes are documented as part of the EMS, which is integrated within the existing quality system. The controls for the monitoring and measuring of the EMS system remain as per the existing management reporting system. The quality manager reporting on quality issues is also responsible for reporting on environmental and health and safety issues. The two main areas of the quality manager's reporting focus are firstly, the reporting of best practice in the manufacturing process and secondly, reporting on environmental issues or pressures.

The best practice is the main objective for Organization A and considerable detail is given to manufacturing best practice and to the environmental requirements that ensure best practice. Due to the lack of US environmental legislation requiring environmental management change, the organization is considered to be an environmentally low-risk organization. There is at present no pressure from the States for an EMS standard. The implementation of the ISO 14001 system is purely to improve manufacturing efficiency, that is less waste, less energy usage, improved design and manufacturing processes. As a consequence the EMS is considered by the environmental manager to be 'invisible to a lot of people and not applicable to the whole organization'.

Available resources

Resources were limited for the implementation of the EMS. Senior managers had agreed that a budget was available but its size was never specified. Separate cases were made to senior management for the level and timing of expenditure. 'Sometimes', said the environmental manager, 'cases were approved and sometimes not. The two managers tasked with the implementation of the system also had full-time, often conflicting, operational jobs. This is not a workable arrangement.'

To assist with the increased workload, a dedicated person was to be seconded to the manufacturing department for six months to write procedures and work instructions. After three months, nobody had been identified to fill the position, the reason being that they could not find anyone who was suitably qualified. Suggestions had been put forward by the two incumbent managers to use a postgraduate student, familiar with EMSs, at the dissertation stage in exchange for some work experience.

Innovation

There are three main business goals for Organization A: first, to grow the organization, second, to regenerate existing products and finally to appeal to a younger market. The EMS would become a subset of the first business objective: to grow the organization as the emphasis is to grow the organization on environmentally sound products. The growth objective also includes developing and adopting innovative ideas in plant technology to make processes more efficient and less harmful to the environment – 'there is certainly considerable management commitment to adopting new technology', says the environmental manager.

Communication

Communication of the organization's objectives is considered to be very good. The organization's intranet is used to provide information on many aspects of its activity including environmental and quality issues. The intranet is plant-wide and delivers the outcomes of quarterly business meetings and quarterly newsletters including those items that promote environmental development to all staff levels. Spreadsheets are posted showing the progress of many waste collection and energy-saving schemes.

Credible plans

The plans for the introduction of the EMS are considered credible in as much as they 'affect a minimum amount of people and there are minor high risk activities'. Incentives for their achievement are considered adequate. The environment manager's view is that 'we do not get a lot of hassle from third parties seeking environmental information, the implementation of the EMS is driven by the manufacturing unit and there is no demand or pressure from other areas of the organization or from the general public'.

Follow-up interview

The person at Organization A was originally interviewed, and its EMS assessed, when there were strong economic conditions in their sector. An opportunity arose to interview the organization following a downturn in this industry sector. It was felt that a follow-up interview would provide insight into whether their EMS model remained consistent in negative as well as positive economic conditions.

The interview was undertaken by telephone with the same person; the environmental manager. The interview was unstructured in that there was not a list of questions to act as a guide. The interviewee was asked to recall the original interview conversation and to compare the situation then and to comment on the aspects of the economic downturn. The participant was asked to relate the economic changes that the organization was currently experiencing and to identify any factors that were occurring that were having an effect on the EMS.

Organization A was being affected by the downturn and recently management had announced that 235 of their 1,750 staff members were to be made redundant. For Organization A, this was considered a 'drastic cutback'. A key directive from senior management asked staff to 'do what you can but spend no money'. Another senior management directive stated that the implementation of the EMS would be shelved for one year. There was no budget allocation for EMS implementation.

The senior management team of Organization A held to the notion that their business activities did not pose a high environmental risk. There was no pressure from customers and the message from the corporate office in the USA was that there was no environmental management pressure as long as the system was not ignored completely. The waste recycling programmes that had previously been set up remained in place and a system audit would be completed. The plan, identified in the first interview, to employ a postgraduate student to write the ISO 14001 procedures has also been delayed for 12 months. The new target for achieving ISO 14001 certification, is now two years later than first anticipated.

The key directive from the USA was that Organization A and its existing product range had achieved market maturity and that the priority for all available resources was the development of new products. This, together with the loss and movement of people, had resulted in a loss of commitment by departments to the EMS. It was anticipated that when the implementation of the EMS was activated again in the following year there would have to be new environmental awareness and training programmes undertaken.

Devoid EMS profile

By selecting the key characteristics of Organization A's case study, the following profile (Table 5.2) is that of an organization which is devoid of an EMS.

Isolated EMS model

The characteristics of an organization with an isolated EMS model are shown in Table 5.3. An organization with an isolated EMS profile has an accredited system but it may only apply to one part of the organization, for example manufacturing, or to the whole organization, and a small team has been established in an isolated unit to deal with the environmental compliance issues.

TABLE 5.2 Devoid profile of Organization A

Characteristics of Organization A: Devoid EMS	No accredited EMS.
Organizational drivers	Improve efficiency.
	Minimize environmental risk.
	To remain competitive.
Operational advantages	Communications believed to be good.
Operational disadvantages	High risk of heavy fines.
	Adverse publicity.
	Lack of shareholder investment.
	Loss of credibility.
	High investment risk.
	Loss of competitive advantage.
	Loss of market share.
	Not accepted on many major supply chains.
	High risk of environmental incident.
Organizational barriers	Management commitment limited.
	Organizational structure considered to be irrelevant.
	System integration limited.
	Available resources very limited including trained staff.
	Innovation wanted but not investigated.
	Credible plans not formulated.

TABLE 5.3 Isolated EMS model

EMS model	Isolated.
Organizational profile	The organization has an accredited EMS.
	The EMS applies to only a part of the organization.
	A small unit or team has been established to deal with environmental compliance.
Operational advantages	Low cost option focused only on compliance.
Operational disadvantages	Low priority of environmental issues.
	Loss of efficiency.
	Fire-fighting environmental incidents.
Organizational barriers	Communication.
	Management commitment.
	Available resources.
	Incentives and controls.
	Company culture.
	Credible plans.
	Management style.
	Organizational structure.

An isolated EMS, by its very nature, is separated from many of the daily operations of the organization. As a consequence there is a lack of communication between the environmental team trying to raise levels of awareness and having input into operational decisions with other managers. The limited resources used to establish the unit send messages to staff members and managers that there is a lack of commitment from senior managers and directors and that environmental issues have low priority against operational requirements.

Case study: Organization B

Background

Organization B commenced business 100 years ago by entering the communications equipment market. In 1950, transistor-oriented research and development led to the beginning of Organization B's semiconductor business. It entered the computer market in 1954 with the start of an extensive computer-related research and development programme. Its growth today has been achieved through the formation of strategic alliances with other successful organizations to provide a range of high specification technology products and services.

The organization's main products and services include the provision of internet business solutions and the manufacture and sale of computers, communications equipment, electronic devices and software. Organization B has a plant in Europe, established in 1981, and after three development phases has a facility of

500,000 square metres and employs approximately 1,500 people in semiconductor manufacturing.

Interview

The environmental manager was interviewed for this case study. He has been with the organization for 16 years and involved with the EMS for four years. ISO 14001 is the EMS being used and certification was achieved three years ago. He was involved with the ISO 14001 system during implementation and certification. He reports directly to the General Manager of Organization B.

Incentives and organizational strategy

The two main drivers for the introduction of the EMS were first, responding to customer and public demands and second, a directive from corporate office requested that all plants worldwide were to be certified to ISO 14001. The corporate office has assumed responsibility for the administration and auditing of the EMS on all sites worldwide within the last 18 months.

Culture and management commitment

The corporate culture is considered conducive to a strong commitment to environmental, health and safety and quality issues. The minimization of environmental risk is one of the main corporate objectives and all levels of management agree and support these main objectives. The three management systems of environment, health and safety and quality are important to the daily manufacturing process. Managers are committed to small activity groups, similar to 'quality circles', to ensure that the standards of the three systems are maintained.

Integrated systems

At present there are three separate management systems: environment, health and safety and quality. Discussions are ongoing as to the benefits of an integrated environment and health and safety management system. The current preference is to merge two and keep quality as a separate system. The reasoning is that the quality system is so intensive and technical because of the nature of semiconductor manufacture that it should not be diluted. Over time, the expectancy is that the three systems will eventually merge into one due to the ongoing merging of ISO standard requirements.

Strategy complexity

As the three systems currently operate as separate systems, there is a degree of complexity as a result of this. The creation of complexity is not considered to be too disruptive as long as each system is part of daily operational activity and the administration of the three systems is similar – the environmental manager admits that the 'technical complexity remains with the quality system but this is the nature of our business. There is often cross-over of system operations but the environmental function is clearly understood by everyone'.

Communication

To ensure this clear understanding, communication is deemed important and there are a number of training programmes and briefing meetings that managers and staff can feed into. There are adequate resources for the development of existing and new communication projects. Although identified as being important for environmental communication and awareness programmes, information technology is little used at present. Environmental documents and procedures are kept in electronic format although not everyone has access to them. An environmental website is currently being developed but is not yet in operation.

Incentives

The environmental manager believes that, despite the lack of financial incentives, existing staff recognition programmes provide sufficient incentives for staff and managers to achieve targets. The organization's newsletter provides one outlet where people can be recognized for achieving environmental targets or identifying innovative ideas. On the subject of innovation he suggests that 'many of the innovative ideas that have emerged to benefit the organization have done so as a consequence of having an EMS. Without it I am sure many would not have emerged; the EMS is a driver for innovation.'

Control

While head office takes a broad product based approach, device manufacturing varies from site to site and it is the responsibility of each site to be more focused on the detail of the environmental issues that emerge and the different types of devices and manufacturing processes used.

The environmental manager believes that as he reports directly to the managing director then necessary resources are more easily achieved. Environmental planning also has greater credibility as environmental and operational requirements have a project focus. Resources are allocated to meet all of the projects' requirements and environmental issues are included in this. Control is strong as the environmental plans are reviewed six monthly as part of the main auditing structure. At present individual sites do their own auditing, these are done on a monthly basis and the managing director is a regular participant. He is kept fully informed and if additional resources are required he is well placed to aid further commitment.

Operational transformation

When asked to identify two factors that would improve or impede the existing EMS the environmental manager felt that significant improvement has been made since the EMS has become integrated with daily operational activities, and he felt that this has helped the organization to be more transparent when environmental information had been requested by customers and the general public.

The second factor identified was seen as impeding the EMS. The lack of environmental awareness and the selling of the EMS's benefits to particular managers impeded the development of the EMS. He states, 'in the early days of EMS implementation this was considered important and during the implementation process it was a benefit, but what seems to have been forgotten is that these issues are still important particularly as personnel, operating processes and corporate objectives and targets change.'

Resources

Like all organizations, Organization B has faced business difficulties. To date it had not announced any redundancies but management were watching the market forces around the electronics industry closely. Four people make up the EMS team and they are still in post. The environmental manager suggests that 'while there is a restriction on expenditure there is a reluctance to interfere with the EMS because it is fundamental to the whole manufacturing process. If the economic conditions persist the rate of continual improvement may slow but the integrity of the systems would remain.'

Isolated EMS profile

An isolated EMS (Table 5.4) is separated from other quality and health and safety systems. Individual project groups' 'quality circles' are responsible for environmental considerations. There is a direct reporting line to the managing director.

Devolved EMS model

An organization operating with a devolved EMS has an accredited system that pervades the whole organization and is part of daily operational activities. It is likely that the manager responsible for environmental management is also responsible for health and safety and possibly quality. The responsible manager plays a key administrative role by communicating progress to senior management by way of periodic reports and ensuring that training and awareness programmes are made available for staff use. Contractors responsible for specific waste recycling tasks and energy efficiency monitoring also report to the organization's environmental manager.

The key operational advantage that a devolved system offers is one of flexibility (Table 5.5). The use of contract staff can aid this flexibility by hiring or downsizing

TABLE 5.4 Isolated EMS profile of Organization B

Characteristics of Organization B: Isolated EMS	The organization has an accredited EMS.
Organizational drivers	Minimize environmental risk.
	Market and customer pressures.
Operational advantages	Greater environmental openness with customers and regulatory bodies.
	New project development includes potential environmental impact assessment.
Operational disadvantages	Lack of environmental awareness.
	Degree of environmental prioritizing behind operational needs.
	Environmental resources dependent upon project approval.
Organizational barriers	Administrative complexity of three independent systems.
	Lack of adequate IT structure.
	Available resources determined on a project-to-project basis.
	Lack of management commitment.

TABLE 5.5 Devolved EMS model

EMS model	Devolved EMS.
Organizational profile	The organization has an accredited EMS.
	Key manager provides an administration support function for contractor activities.
Operational advantages	Increased downsizing flexibility.
Operational disadvantages	High cost option.
	Loss of environmental knowledge and expertise.
Organizational barriers	Gaps in communication.
	Management style.

whenever the economic situations dictate. The high hourly rates of engagement are considered an acceptable price to pay for downsizing flexibility. While the administration of the system is handled by the organization there is a danger that downsizing operations cause the loss of environmental knowledge and experience of operational processes.

This style of management is seen as a key organizational barrier to the development of an EMS as the focus is on short-term costs and not on long-term environmental improvement. Communication is also considered a barrier as contractors constantly change and communications can be interrupted. A 'them' and 'us' attitude between organization staff and contractors can lead to informal demarcation lines and grey areas of responsibility.

Case study: Organization C

Background

In 1999, Organization C split from its parent company and was listed as a public organization on the New York Stock Exchange. Organization C operates four businesses: test and measurement, semiconductor products, healthcare solutions and chemical analysis, supported by a central laboratory. Its businesses apply measurement technologies to develop products that sense, analyse, display and communicate data. Organization C's customers include many of the world's leading high-technology organizations, which rely on Organization C's products and services to make them more profitable and competitive, from research and development through to manufacturing, installation and maintenance. Organization C enables its customers to speed their time to market and to achieve volume production and high quality, precision manufacturing.

A key driver for Organization C's products and services is the pervasive transformation from analogue to digital technology. Because digital technologies require greater degrees of precision, and rely more on miniature circuitry than analogue, the role of test and measurement is 'mission critical' for the rapid commercialization of reliable internet-age products.

The organization has 48,000 staff and facilities in more than 40 countries and serves market leading customers in over 120 countries. Major product development and manufacturing sites are located in the United States, China, Germany, Japan, Malaysia, Singapore, Australia and Europe. More than half of the organization's net revenue is derived from outside the United States.

Based in Europe, Organization C operates in the telecommunications testing industry sector providing industrial test equipment to major telecommunication product manufacturers. It employs 1,200 people in a manufacturing site of approximately 300,000 square metres. The site separated from its parent company in 2000, having originally been the testing facility for the parent company.

Interview

The person interviewed was the product steward manager who is responsible for ensuring that a high percentage of recyclable material is designed into Organization C's products. He has been with the organization for three years and has been involved with ISO 14001 EMS for all of this time and reports to the manufacturing engineering manager.

He sees the two main drivers for the organization having an EMS as first, to remain competitive and second, to reduce environmental risk. These drivers were initially part of the parent company's corporate policy. Organization C was originally the 'beta' test site for the introduction of ISO 14001. After the business separation, the EMS remained with the new business; as the same operations remained, only the name changed.

Management commitment and management style

Despite the separation, senior management remained committed to the EMS and applied a corporate directive to other sites to introduce ISO 14001. At present the commitment from senior management to reduce environmental risk is considered strong. The organizational structure is a lot flatter than the parent company's and more devolved. During the time Organization C was part of the parent company it was more difficult to get senior management to support environmental management issues due to the increased number of organizational layers. A consequence of this structure was that there were more political considerations and there were more business/environment trade-offs.

The product steward manager believes that with Organization C, the management style is less autocratic than the parent company and seems more suited to cope with a higher level of environmental risk. However, environmental management is never a high priority because of the nature of the business. He points to the fact that:

> because we [Organization C] produce low volume, high quality industrial products and not commercial products, we have no consumer pressure groups driving environmental standards. We do, however, listen to our major product purchasing clients who expect us to adhere to their environmental standards, but here there is not the same pressure.

> The existing corporate culture does aid the development of the EMS, but due to the current economic circumstances it is not top priority. It is still very much part of our daily operational process and product design but just as with quality and health and safety the speed of development is slow.

Prior to the industry downturn senior management were more committed to addressing the development of the EMS. The product steward manager accepts that the main reason for this would have been as a result of the split from the parent company. He adds that 'much environmental expertise had been lost through the departure of personnel and much EMS knowledge and experience was, and is still required to incorporate the products and services of [Organization C] as opposed to [the parent organization's] original computer manufacturing activities.'

Resources

He believes that 'the EMS development commitment still exists, across other sites in Europe and America too, but it is the current lack of available resources that slows the development of the EMS.' Initially, with the parent company, strong economic periods signalled significant available resources. Following the split from the parent company, significant resources were also made available for the revamping of the ISO 14001 standard to Organization Cs' products. Now, during the sector downturn, resources are not so abundant but this applies across the organization, not just to the EMS. 'Having an integrated system means that if daily operational activity decreases everything that is associated with it also decreases.'

The product steward manager believes that ICT plays a large part in the system integration process because the guidelines, documents and procedures for the EMS and the other systems are based on an almost paperless system. The signing of documents and procedures is also done electronically. He says 'our passive approval documentation process means that the document or procedure is seen by everyone and is accepted unless an objection is raised. This speeds everything up, although caution does have to be exercised.'

System integration

All of the management information systems such as environment, quality and health and safety are integrated into a single system. 'This system was planned by senior management, following the split from [the parent company], as a method to improve efficiency and reduce operating costs.' It was considered that as there was a database of legislation affecting the organization, it did not matter whether it was related to environment or health and safety, it was relevant to how the product was designed and manufactured. Quality is also built into the design and manufacturing of the products so it is considered that an integrated approach works.

Strategic complexity

There is little complexity as plans, procedures and controls for the EMS are based on those guidelines provided as part of the ISO 14001 and ISO 9001 standards currently being applied. He adds, 'we do not want to reinvent the wheel. Communication plays a big part in reducing complexity; following the "split" it was essential to ensure the organization operated effectively as a separate entity. Within [Organization C] there is currently an ongoing project to integrate all databases into one large, centralized database – at times we can suffer from having too much information.'

Incentives

For the majority of staff there are no financial incentives to reward them for an EMS that operates efficiently and develops continuously. Recognition for new ideas for waste reduction or energy efficiency schemes is given in the corporate newsletter. Some senior and middle managers do have some financial incentives for environmental improvement but these would have been individually negotiated into their contracts of employment. In the current economic climate it is felt that financial incentives would be difficult to justify, particularly as all staff, managers included, have recently been asked to consider a 10 per cent salary reduction in order to avoid redundancies.

The two most difficult areas for the EMS at present are first, the lack of EMS champions within the organization that have influence among senior managers and second, the lack of short-term environmental benefits that can be traded off with senior management for investment into long-term environmental improvements. He suggests that 'the short business cycles from product development to obsolescence and trying to sell the long-term benefits of proposed environmental product improvements can, at times, be difficult to reconcile.'

Innovation

It is recognized that there is a need for innovation within the EMS but at present, possibly because of the split, the most innovative ideas are being used to put forward arguments to obtain additional investment for long-term environmental improvements.

Current economic situation

Organization C has recently put a proposal to the workforce to accept a 10 per cent pay reduction to avoid any redundancies. As major clients within the electronics and telecommunications sector are continuing redundancy programmes Organization C is monitoring the situation closely before making any further announcements.

Devolved EMS profile

Contractors responsible for carrying out environmental duties, together with the organizational manager, have multiple tasks and have a reporting administrative function to senior managers, and a communications function to all staff members. EMS is integrated with health and safety and quality systems and ICT plays an important part in EMS administration (Table 5.6).

Case study: Organization D

Background

With more than 12 million products sold since the organization was founded in 1972, Organization D is recognized as a leader among personal computer (PC) peripherals (printers, and multifunction devices) with operations in the USA, Canada and Latin America.

Organization D's parent is a $5.6 billion multinational corporation and a world leader in information processing systems, telecommunications and electronics. It is ranked first among printer manufacturers in North America with a 37 per cent market share, and is ranked second among vendors in the laser printer category with a 6.5 per cent market share (Dataquest, 1999).

It has one of the broadest product lines in the industry, marketed under the brand name and including graphic imaging technologies of digital light emitter displays (LEDs) and colour imaging. Organization D is the main supplier for Europe for

TABLE 5.6 Devolved EMS profile of Organization C

Characteristics of Organization C: Devolved EMS	The organization has an accredited EMS.
Organizational drivers	Minimize environmental risk.
	To remain competitive.
Operational advantages	Standardized system administration – one information database.
	Flexibility to up-size and down-size quickly.
	Innovation comes to the fore.
	Organization has flatter structure.
	Managers have more hands-on role, closer to daily operations.
Operational disadvantages	High cost of operational flexibility.
	Potential for loss of knowledge and experience.
	Lack of EMS champions.
	Lack of long-term environmental investment.
Organizational barriers	Lack of available resources.
	Lack of communications.
	Lack of management commitment.
	Speed of organizational change.

these products and employed 900 people on a manufacturing site of 250,000 square metres.

Interview

The person interviewed was the environmental manager. He had been with the organization for 11 years and involved with the EMS from the implementation process that began in 1997 and culminated with final certification in April 1998. He reports to the Operations Support General Manager who is responsible for the manufacturing support services which includes the environmental, quality and health and safety systems.

The main drivers for the organization having an EMS have changed over a number of years. In 1992 there were two main drivers; first, adhering to existing environmental legislation and second, market pressure. In 1996 a third driver was added by way of the implementation of a corporate directive that required all manufacturing sites to have an accredited EMS based on the ISO 14001 standard.

This corporate directive ensured commitment from senior management. Prior to 1996, senior management commitment for EMSs was sporadic and varied from site to site. Environmental issues prior to 1996 were considered to be low priority and not part of the corporate culture. This acceptance of environmental responsibility for each site mirrored the cultures of environmental responsibility found in Japan, America and Europe.

The plant in Europe began with three distinct management systems: environment, quality and health and safety. The objective for the plant was to eventually have one fully integrated management system incorporating all three systems. The roles of the environment, quality, health and safety managers eventually merged to form a team of individuals that were responsible for all site facilities management.

This restructuring has the effect of flattening the organizational structure and current plans are designed to continue this process, aiding the development of one fully integrated system. The environmental manager considers that this is a good move as 'managers are getting closer to the operational "front-line" and are becoming more aware of the daily operational and environmental requirements.'

Integrated systems

Effective administration of an integrated system is also important. Most of the environmental documents and procedures are in electronic format but that is not sufficient according to the environmental manager; he feels that 'to enable a fully integrated system a more focused approach is required on the use of ICT.'

At present the existing word documents and intranet services are considered wholly inadequate for the proposed integrated system. Software packages do exist to aid an integrated system and a review is underway to determine whether existing 'off-the-shelf' software packages are suitable or whether a customized software package needs to be written. He believes that 'managers accept that there are many similarities between the three management systems, which is a big step forward. The use of software for an integrated management system only works if the administration of the system is standardized. The integration of an EMS also requires the "cross-training" of auditors to audit all systems.'

Credible plans

When asked about how credible the plans were for the move to an integrated environmental system, he remarked, 'Initially, for the introduction of the EMS we followed the ISO guidelines for 14001 but as we have developed the system so further plans have been presented by management. As we are not sure how to progress to a fully integrated system, these plans are being written as we speak and they have to be approved by senior management before they are accepted. There may be a possibility that this site takes on the environmental responsibilities for all European sites and we will have to amend our EMS plans again.' The feeling among managers until just recently has been that an integrated environment, health and safety and quality system was not needed. He feels that the three systems could exist quite successfully separately. However, he says, 'The recent down-turn in the electronics industry is leading to a rapid move to a more integrated system to aid flexibility and reduce costs'. He believes that 'Integration has been somewhat thrust upon us but, on the positive side, it provides an opportunity to think about the systems and reconsider their efficiency and effectiveness. We anticipate improvement from a long hard look at what the systems are there to do and the roles of personnel within the systems.'

Communication

These continual changes make communication an important issue within the organization. Frequent change puts pressure on the internal communications structure to keep all staff members informed. This communication process will be strained further with the current economic changes as some staff members will be made redundant, some will have new responsibilities and a greater number of contractors will be utilized as more environmental services are outsourced.

Resources

These changes also put a strain on available resources. Initially, there were sufficient finances and people to support the operation of the environmental, quality and health and safety systems but now as a result of the restructuring a number of people have been lost from the EMS.

> The energy manager and the facilities manager have been made redundant and most of their responsibilities have been given to me along with a, still to be confirmed, title of 'Site Facilities Manager'. The remaining environmental personnel are mostly contract personnel. Resources at the moment are scare and each environmental project has to be presented and approved on its merits by senior management.

It was felt that the increased use of contractors as opposed to staff members would allow the organization to be more flexible in its operations as these could be turned on and off depending on the economic circumstances. The downside was seen to be that the constant change of personnel would be expensive and lead to a loss of environmental knowledge, experience and responsibility. The constant process of change would lead to the continual reinvention of the environmental management wheel but not necessarily continuous improvement – the ethos of the ISO 14001 standard. It is considered that the greater the rate of change the harder it is to build awareness and maintain environmental commitment. Conversely, if change is too slow apathy can creep in. The view is that some change is good for continual improvement but a high rate of change can create chaos.

Current changes to the organization include moving the manufacturing facility to Thailand, reducing staff numbers from 900 to 500. It is thought that everything would have to be realigned with the reduced business requirements. This includes procedures and systems. The organization would have to go through the awareness cycle again as some people would have left and others would have changed responsibilities. This would affect all staff levels.

Losing people from the environmental and health and safety functions has added to the workload. The environmental manager says that 'this requires us to devolve some functions to external consultants, particularly auditing. People have been demoted, promoted and the whole organizational structure has been flattened. Middle management has been stripped out. Environmental budgets remain but they have been reduced along with everything else. Problems are being experienced even now with devolving responsibility and levels of professionalism.' It is perceived that operating with a flat structure means that the main focus is on cost. The environmental principle of 'best practical environmental option' (BPEO) is used to assess the viability of every project. He says, ' we are now working with financial targets as opposed to environmental targets. Environmental priorities are low priority by necessity. We work within economic constraints where the emphasis is on balance. Budgets for all systems are required to do more and go further.'

With the loss of key personnel the management of three separate systems would be a complex mix, and it is hoped that the streamlining of the three systems into one works to alleviate the complexity. He feels he is being forced to consider what the organization is trying to achieve and therefore cut out the dead wood. He says that

> innovation is beginning to come to the fore, working on the assumption that 'necessity is the mother of invention' the lack of resources is forcing us to consider innovative ways of overcoming environmental issues. One recent example was the reduction of the high cost of disposing of condensate trade effluent; a new separation system was designed and built to separate oil and waste and divert to soak-a-ways. This averted an expensive draining system that was originally planned to cope with the produced condensate.

Innovation

It is thought that innovation could play a bigger part in the EMS development process if there was a greater emphasis on incentives and rewards for new ideas. At present there are no financial incentives to aid the development of new environmental ideas; the only process of acknowledgement staff have is the site newsletter. Some incentive is provided at the induction process where staff members and contractors are advised of waste recycling and energy efficiency schemes. Monthly management meetings also exist to give managers the opportunity to encourage existing staff in the pursuit of environmental targets.

It is perceived by management that there is real danger with the constant change to the numbers of staff and increased operational responsibility, that environmental objectives are gradually forgotten about. As well as the strong emphasis on the induction process, individual staff appraisal programmes act as a sufficient control system to ensure that a high level of environmental awareness is maintained. Regular environmental management reviews are also considered to be of assistance.

It is hoped that, given current economic conditions, the frequency of these two important control mechanisms remains and is not reduced.

It is not considered that the EMS is being reduced in itself, as environmental issues are part of the daily activities of the organization and the overall operational budget allocation is being downsized together with other management systems and operational requirements. He argues that, 'The organization is striving for balance rather than excellence; after all, we are in competition with our own sites as well as those of our competitors.'

Devolved EMS profile

As with Organization C, contractors are responsible for carrying out environmental duties and the organizational manager has multiple tasks and has a reporting and administrative function to senior managers and a communications function to all staff. Integrated EMS with health and safety and quality systems and IT plays an important part in EMS administration (Table 5.7).

EMS profile – devolved EMS

Integrated EMS model

An organization with an integrated EMS operates with one system that incorporates their environmental, health and safety and quality systems. The

TABLE 5.7 Isolated EMS profile of Organization D

Characteristics of Organization D: Devolved EMS	The organization has an accredited EMS.
Organizational drivers	Minimize environmental risk.
	Market and customer pressure for environmentally sound products and services.
Operational advantages	Organization has flatter structure and managers have more hands-on role and are closer to daily operations.
	Flexibility to up-size and down-size quickly.
	Innovation comes to the fore.
	Standardized system administration.
Operational disadvantages	High cost of operational flexibility.
	Loss of knowledge and experience.
	Less environmental control.
Organizational barriers	Lack of available resources.
	Lack of communications.
	Lack of management commitment.
	Speed of organizational change.

single system, often renamed an EHS system, is designed to operate with the same documents and procedures and 'cross-skilled' auditors to monitor the system (Table 5.8).

The key operational benefit of an integrated EMS is that it becomes part of daily operational activities. It is a system that is part of the organization's goals and objectives and as such it has the commitment of all managers and directors. Monitoring daily operational activities gives management early warning of the development of potential environmental incidents. The development of an integrated EMS gives an organization an opportunity to develop a culture of continuous improvement.

An operational disadvantage that can occur is that the establishment of a corporate centralized auditing unit can result in loss of intimate environmental knowledge of site-specific manufacturing processes. Additionally, an audit programme may structure the required audits to be undertaken every two or three months. Some site-specific processes may require more regular audits and as such the early identification of potential environmental incidents may go undetected until they fully develop.

There seem to be few key organizational barriers that can occur with an integrated EMS; however, communication is still important to maintain staff awareness and participation. Many environmental training and awareness programmes exist within the organization but difficulties can result with the lack of staff member participation. The daily interaction of environmental issues with operational activities may render them uninteresting over time. Communication is the key to keeping the environmental training and awareness programmes fresh in the minds of staff. This may involve periodically changing the scope of the programmes or introducing competitions and offering cash prizes or holidays.

TABLE 5.8 Integrated EMS model

EMS model	Integrated EMS.
Organizational profile	The organization has an accredited EMS.
	EMS is customized to fit with H and S and quality systems to form one system (e.g. EHS).
Operational advantages	Part of daily operational activities.
	Early warning of potential incidents.
	Culture of continuous improvement.
Operational disadvantages	Centralized auditing process can lead to loss of site process knowledge.
Organizational barriers	Lack of employee participation.
	Communication.
	Management style.

Although managers and directors all agree to support an EMS, there is an element of interpretation. There can be many management styles within an organization and these different styles may interpret environmental management in different ways. The outcome may be disagreement on the operational methods to be used to achieve the environmental objectives.

Case study: Organization E

Background

The first product of Organization E was a 'battery eliminator' allowing consumers to operate radios directly from household current instead of the batteries supplied with early models. In the 1930s the organization successfully commercialized car radios. During this period, the organization also established home radio and police radio departments, instituted pioneering personnel programmes and began national advertising.

The decade of the 1940s also saw it begin government work and open a research laboratory in Phoenix, Arizona, to explore solid-state electronics. By the 1960s, Organization E was a leader in military, space and commercial communications, had built its first semiconductor facility and was a growing manufacturer of consumer electronics.

Organization E expanded into international markets in the 1960s and began shifting its focus away from consumer electronics. The colour television receiver business was sold in the mid-1970s, allowing Organization E to concentrate its energies on high-technology markets in commercial, industrial and government fields. Today, Organization E is using the power of wireless, broadband and the internet to deliver embedded chip system level and end-to-end network communication solutions for the individual, work team, vehicle and home.

Control

Organization E is an American-owned organization. All management systems are controlled from a centralized corporate office and a dedicated person, the Crisis Management Vice-President, heads a separate risk management department responsible for monitoring the environmental health and safety system (EHS) worldwide. Environmental risk is taken seriously by the organization, and it is one of the six corporate objectives on the high priority list.

The controls for the efficient running of the EHS are devised by corporate office based on the guidelines and principles laid down by the ISO standards. Managers accept the organizational objectives and targets of the business plan and consequently accept the controls that go with achieving these objectives. There are disagreements at times because managers may interpret environmental requirements in different ways. There are often intense discussions on what should be environmental goals but these are eventually resolved.

Integrated system

The current EHS is based on ISO 14001 and the health and safety and environment system work as one system. The three main drivers for the introduction of the EMS were first, moral and ethical issues, second, the legal requirements and third, economic – greater savings from improved efficiency for operating processes, waste reduction and energy efficiency.

Having an integrated EHS system as opposed to separate environmental and health and safety systems means there is less operational complexity. The administration of the system, that is, procedures and documentation, is standardized. The corporate auditing team is multi-skilled and audits the EHS system as a whole and not just the environmental or health and safety elements of it. All elements of the EHS system are linked to the business goals and all managers and directors agree these goals. Operational goals are manufacturing related and within these goals the environmental, quality and health and safety targets are clearly specified and accepted by everyone as being part of daily operational activities.

Management commitment and resources

The positioning of a dedicated senior corporate manager for the EMS is seen as necessary to ensure that sufficient resources are made available to support the EMS. A large team of corporate professionals are available to monitor and audit the EHS. Originally each site had their own regional corporate audit team; however, this was closed down and centralized within the 'states' to improve efficiency and standardize auditing activity.

Individual sites still have autonomy in running their own EHS systems but it is the site services manager not the environmental manager who has responsibility for managing the environment, and the health and safety facilities. No one person in the organization is designated 'environment manager'.

The organizational structure supports the development of the EHS, and there is a clear line of reporting and clearly defined environmental management responsibilities throughout the system. As one of the key corporate objectives, environmental management has the support of all managers and directors. It is accepted that EHS support is not wholly prescriptive and that managers and directors do exercise some element of interpretation as to the level of support required.

There is a general view that the operation and development of the EHS is well resourced, although at present there is a requirement for a further two environmental and two health and safety personnel.

Communication

The biggest issue for the organization as regards the EHS is communication. There are 3,200 staff members within the organization and the majority of them work on a shift basis. There are four shifts and keeping everyone informed is a major task and one that is still not fully effective.

There are continuous EHS training programmes made available to staff to ensure environmental communication and awareness is maintained. All managers commit to the programme by allocating time for all staff to attend the programmes. Monthly meetings are held by small teams from each shift to review objectives and successes, and this information is fed back to each shift. Weekly audits are undertaken across the four shifts to ensure that the communication and awareness activities are integrated into daily operational activities.

It is the integration of the EHS into daily operational activities that makes the environmental management development plans credible and acceptable. The facilities manager suggested that an integrated approach ensured that 'staff member environmental awareness and responsibilities were not additional burdens but were part of a normal working day'.

People are encouraged to approach management to propose change. Open days and organization newsletters are used as vehicles to assist with opportunities to express views for environmental improvement.

Innovation

Innovation is seen as an important part of keeping the EMS fresh. It is also the hardest area to implement. Environmental goals and objectives are used to introduce innovative ideas into the system; the facilities manager states that for him

> the hard part is getting people to participate; awareness levels are high but the hard part is getting people involved; constant communication is required to feed back the successes of innovative ideas. We have a saying here, 'everyone comes here with two arms and two legs and they are entitled to go home with the same'; it is our responsibility to make sure this happens. We have to innovate to keep everything fresh.

Incentives

While continual operational cost savings are beneficial, making environmental issues part of daily operational activities provides an early warning system for identifying potential environmental incidents. 'We have had no prosecutions on site to-date,' says the site facilities manager and adds, 'That's not to say there haven't been incidents, there have, but they were identified early and dealt with.'

The current incentive and reward schemes act as big incentives towards continuous environmental improvement. There is keen competition between sites on environmental performance outcomes. Trips to various countries are on offer to those individuals or teams that excel in environmental or health and safety improvements. A dedicated corporate team affectionately referred to as the 'MET Office' (Management Ergonomic Team) is a multidisciplined team of 12 members who assess improvements and recommend the rewards. Direct financial incentives are also available and are assessed through the staff appraisal system.

Barriers

The two key factors the site facilities manager identifies as impeding the development of the EHS are first, the centralized auditing team have overall expertise in monitoring the EHS but at site level there is a tendency to generalize environmental issues and therefore a danger of intimate knowledge of site processes being diminished, or lost, thereby increasing the risk of an environmental incident. Second, inconsistent commitment to internal EHS public relations can have an impact. The site facilities manager believes that 'this fluctuates between good and could be better, but it is important to maintain high awareness and training levels to ensure continued management and staff commitment.' He notes that the organization recognizes that local community development is crucial to the continued success of the organization. High commitment is given to free local environmental training and education to pupils and students. Due to the current economic uncertainty some of the educational projects are on hold but the task of reassuring the community is at present top priority.

Integrated EMS profile

By selecting the key characteristics of Organization E's case study, the following profile is that of an organization with an integrated EMS (Table 5.9). The integrated system referred to as the EHS system is part of the organization's daily operational activity. It has one management information system with corporate environmental

TABLE 5.9 Integrated EMS profile of Organization E

Characteristics of Organization E: Integrated EMS	The organization has an accredited EMS.
Organizational drivers	To remain competitive.
	Improve efficiency.
	Minimize environmental risk.
Operational advantages	Early warning of potential environmental incidents.
	Improved efficiency.
Operational disadvantages	Size of company and shift work slows environmental communication.
	Decentralized auditing system which causes the loss of site process environmental knowledge.
Organizational barriers	Slow communication channels.
	Lack of employee participation.
	Management interpretations of environmental requirements.
	Lack of IT development.

objectives, which is part of a business plan agreed by all directors and managers. There are high incentive and reward schemes for new environmental initiatives and the integrated system expands and contracts with the level of business activity.

Case study: Organization F

Interview

Organization F's European manufacturing plant was the first site to achieve an accredited EMS. It achieved certification for ISO 14001 in August 1998 after a two-year implementation process. The site facilities manager (interviewee) had not installed the existing ISO 14001 system but had operated the system for the 20 months he has been with the organization. Previously he was the environmental manager with another computer-based organization for ten years and had been involved with implementing their ISO 14001 system. He reports to the Environmental, Health and Safety and Security Director.

There were two main organizational drivers for the pursuit of this environmental standard. The first was to attract and retain large commercial customers by demonstrating an environmental commitment demanded by large international organizations. The second was to achieve parity with a competitor that had achieved the standard in 1995, and to stay ahead of other competitors that had yet to achieve the standard. Since 1995, ISO 14001 had been classified as a major discriminator (a significant marketing advantage or disadvantage) by all electronic Original Equipment Manufacturers (OEMs).

The organization addressed its environmental objective in the same way as its other business standards (62 in total). All of the organization's business standards were built into the corporate business plan. The directors and managers accept, or buy-in to, the achievement of the business plan. The practice is cost-effective as all directors and managers agree the business plan objectives.

Resources

Budgets are made available and the directors and managers are aware that they must allocate resources to address environmental considerations as and when the situation arises. Each operational facility within Organization F is assessed on these business standards, although the European-based facilities are the only ones that have so far achieved the environmental standards. The assumption is that the facility manufactures products for Europe and an environmental standard is required if it is to compete successfully in an international arena. The EHS manager stated that Organization F was striving for a worldwide (all international sites) EMS accreditation system. The international organization IBM was presented by management as an example of a large electronics organization that began with individually certified EMS sites before moving to worldwide EMS accreditation. This was considered to be an acceptable way forward for Organization F.

Culture

The culture of the organization is 'States driven' and everyone from director to shop-floor operator buys-in to the achievement of set targets and objectives and when these are realized they are all rewarded for their efforts. Much of the environmental improvement planning is based on specific projects such as waste minimization, energy efficiency and material reuse savings. This planning method is seen to have acceptable credibility as managers and staff members are rewarded as the organization reduces its operating costs.

Communication

The achievement of environmental objectives is communicated well to all staff through utilization of internet and intranet systems. Details of efficiency savings are issued in monthly spreadsheets which are posted onto the environmental intranet website. Soft copies of ISO 14001 procedures and work instructions together with quality and health and safety procedures were also available for individual inspection. The website is also made available for all staff to offer environmental improvement suggestions.

Innovation

The generation of innovative ideas is encouraged through departmental 'innovation' sessions and the good ideas are communicated throughout the organization. The environment website is treated as a newsletter and is accessible by all staff through desktop PCs. Keeping control over the number of innovative ideas for environmental improvement was seen by the EHS manager as the main challenge. He found that 'individual enthusiasm sometimes needed to be restrained or better integrated'. Reference was being made to those staff members who unilaterally devised and implemented their own environmental improvement projects outwith ISO standard guidelines, attracting a series of system non-conformance reports.

Strategy complexity

Compared to its main competitor, Organization F management systems are not considered complex in nature as they are treated as one system. All procedures are on the same database and address environmental, quality and health and safety issues within the same procedure. By way of contrast, their main competitor's system was complex; it had three separate systems operating in three different business units with three different environmental managers.

System integration

As the EMS was an integrated part of Organization F's overall business management system this did lead to difficulties with control, particularly when it came to EHS external auditing requirements. The BSI having certified Organization F for ISO 14001 was also tasked with the periodic (annual) audit of the system. One major requirement of keeping the ISO 14001 standard is a formal minute of the annual EMS management review. The fully integrated system operated by Organization F was reviewed by senior managers on a 'business plan objectives achieved' process and not the ISO 14001 specified management review basis. After much discussion Organization F managed to convince the BSI that the annual review of the business plan serves the same function as the management review. This dispensation was also extended to include the ISO 9001 quality standard.

A fully integrated EMS was well received by all managers and directors. One key benefit identified was that the management systems, environmental, quality and health and safety could be audited by one cross-trained corporate auditing team. The corporate auditing was very effective ensuring that periodic audits were carried out as and when required for each system.

One noticeable drawback to a fully integrated management system was that the corporate quality manager (based in the USA) was coordinating the auditing team to do the required environmental audits. The EHS manager said

> it was becoming apparent that there was a lack of in-depth knowledge of specific sites, particularly on particular processes and their potential environmental impact. A site-based auditor, with in-depth knowledge of processes, services and products could control more effectively the frequency and the focus of environmental audits. There is a requirement for site-based EHS managers to drive the EMS and to continually engage and train staff both on a formal and informal basis.

The current structure of the organization, that is, the way that the organization as a whole reports to the USA with one integrated management system and one business focus helps the development and the functioning of the EHS. There was a proposal presented by management to split the site into four main business divisions. The existing site of Organization F is home to four independent business centres that report directly via their own business directors to corporate head office in the USA. It is feared that this change of structure may work to fragment the existing EHS system and create four separate environmental, quality, and health and safety systems.

Management commitment

The hesitancy among managers about the four-business proposal is that as each business director has their own set of operational objectives, as a consequence environmental issues may not carry a high priority as at present. There is also a danger of information being lost between divisions and that available resources are spread too thinly. At present business and environmental trade-offs are minimal due to everyone having the same business plan objectives to be achieved. Four business units under one roof may require greater business and environmental negotiations between the EHS managers and the business unit directors to ensure key environmental issues are taken on-board.

One of the criticisms of the existing EHS system is that bureaucracy is still too high. Although all procedures are stored and referenced on the intranet, physical signatures are still required on 'hard' copies that are often delivered by hand.

Follow-up interview

Interviews were undertaken in Organization F and their EMS assessed, when there were strong economic conditions in the telecommunications and electronics sector. An opportunity arose to interview the organization following a downturn in this industry sector. It was felt that a follow-up interview would provide insight into whether the created model remained consistent in positive as well as negative economic conditions.

The interview was undertaken by telephone with the same person – the site facilities manager. The interview was unstructured in that there was not a list of questions to act as a guide. The interviewee was asked to recall the original interview conversation and to compare the situation then and to comment on the aspects of the economic downturn. The participant was asked to relate the economic changes that the organization was currently experiencing and to identify any factors that were occurring that were having an effect on the EHS.

The interviewee stated that the existing EHS system remained as one integrated unit incorporating environment, quality and health and safety. However, the organization was in the process of making 1,100 of its 3,200 workforce redundant due to the economic downturn. The profit margins for the site that manufactured PCs for the consumer market were now considered to be financially unsustainable, so a decision had been taken to move the manufacturing process to a low-cost country. Organization F operates two sites in Europe; one site manufactures mainframe computer systems and employs 1,500 people. The products made at the second site were highly profitable electronic goods and there were no plans for any redundancies. Both sites have the same EHS system and it remains unchanged.

With the economic downturn Organization F's business model has changed. The emphasis is now on high value-low volume mobile computing devices. To achieve this change fewer staff were required as the 'build' process is less intensive than previous products. The whole business is being downsized. The EHS is also being downsized but in proportion to the whole organization. Key people are being lost in many departments including three in the EHS department. The EHS manager believes that 'as the organization's manufacturing activities shrink so will the EHS and the administrative and reporting responsibilities also remain and adjusted to fit with the resized organization.'

Due to the ongoing redundancies and redeployments, it is envisaged that a new awareness and educational programme needs to be undertaken. The timing of new environmental initiatives is important. At the moment there are many sensitive issues within the organization that are taking priority over everything else. For instance, some departments are due environmental audits but these have been cancelled as many staff are unsure whether they would remain with the organization.

As a result of the downsizing, the main barrier acting against the EHS is that of lack of commitment from staff as well as managers. Managers' attention is taken up with redundancy issues and production cost reductions. Staff attention is taken up with the uncertainty of who is to be made redundant and looking for alternative employment opportunities. This lack of commitment is expected to be short-term – possibly six months.

The speed of change is also resulting in poor communications. As the whole organization is downsizing, so many messages are being circulated, some are being misunderstood and some are not received. The speed of communication cannot keep pace with the speed of organizational change and consequently changes to the EHS. As managers are being given increased responsibility they are required to wear several hats, some with conflicting objectives.

While the administration function, although downsized, remains unaffected, participation from other managers has fallen off, internal audits have been scheduled less frequently and environmental meetings for the time being have been stopped. These are considered short-term symptoms and it is anticipated that the 'normal' functioning of the EHS system resumes when the redundancy program has been completed.

Integrated EMS profile

By selecting the key characteristics of the Organization F case study, the following profile is that of an organization with an integrated EMS (Table 5.10). Like

TABLE 5.10 Integrated EMS profile of Organization F

Characteristics of Organization F: Integrated EMS	The organization has an accredited EMS.
Organizational drivers	Improve efficiency.
	Minimize environmental risk.
	To remain competitive.
Operational advantages	Improved efficiency.
	Early warning of potential environmental incidents.
Operational disadvantages	Size of company and shift work slows environmental communication.
	Decentralized auditing system causing a loss of site process environmental knowledge.
Organizational barriers	Slow communication channels.
	Lack of employee participation.
	Management interpretations of environmental requirements.
	Lack of IT development.

Organization E, this integrated system referred to as an EHS system is part of daily operational activity. It has one management information system with corporate environmental objectives, which is part of the business plan that has been agreed by all directors and managers. There are high incentive and reward schemes for new environmental initiatives and the integrated system expands and contracts with the level of business activity.

Use of 'devoid', 'isolated', 'devolved' or 'integrated' models

The application and implementation of an EMS into any organization invariably produces organizational barriers that act for and against implementation. The discussion through the EMS models in Chapter 4 and the case studies in this chapter suggests that environmental management models can be used as tools to improve environmental performance and assist decision-making in three key areas:

- Organizational EMS development.
- Economic change.
- Competitor discriminator.

Organizational EMS development

The literature and case studies demonstrate that there is consistency in the key organizational drivers that motivate the implementation of EMSs. Based on organizational driver consistency the application of the models in this chapter can provide management with insight into how an EMS fits within an organization, the barriers that can arise from adopted models and the associated operational advantages and disadvantages arising.

To assist management to make informed decisions about environmental performance improvement, it is not enough just to identify the organizational barriers that exist to impede the development of an EMS. The models in Chapters 4 and 5 clearly show the types of barriers that may occur (Chapter 3) and the way they manifest themselves within an organization varies between models and the type and size of organization. It is important for management to understand what causes barriers to occur and the resultant impact upon the system and the organization, so that correct remedial action can be taken. For example, communication is a barrier that can be identified in all models and yet management action to overcome the barrier varies between models. The devoid model, for example, demonstrates that simple communication of environmental information to all staff, if missing, causes a barrier, whereas the integrated model points to the constant refreshing of the existing communication system to maintain levels of awareness and participation and avoid barriers of lethargy and complacency.

Economic change

For organizations that need to react quickly to economic change, the models in Chapters 4 and 5 offer additional management information as to how an organization and its EMS can react in both strong and weak economic climates. For example, organizations that operate in a highly competitive market sector may consider that the implementation of a devolved model offers greater operational flexibility when adjusting to frequently fluctuating market conditions. The high cost of operating with a devolved model may be justified with the need to retain operational flexibility and acceptable profit margins.

Competitor discriminator

Those organizations that seek to be market leaders view environmental management not just as an environmental risk minimization exercise but more as a business growth and competitive advantage opportunity. To this end, adopting an integrated model ensures that environmental issues are treated as all an organization's activities and despite economic growth or contraction, the integrated EMS remains part of daily business activity and grows and contracts accordingly.

Identifying the types of models used by competitors or key suppliers provides management with a useful discriminator to gauge the level of competitive advantage and the likely reactions of these organizations in changing economic conditions. The competitive advantage gained from an effective EMS does begin and end with the type of model utilized but can also extend to the types of models an organizations supply chain demands.

Networked EMS model

The logical next step for those organizations in pursuit of SD objectives and working to deliver products and services for a low carbon economy is to form subnational, national and transnational networks. These networks share the risk of developing new technologies, products and markets and provide access to a shared knowledge, capability and commitment that contributes to long-term environmental, social and economic improvement. This concept is explored further in Chapter 6.

6

ENVIRONMENTAL MANAGEMENT FOR A LOW CARBON ECONOMY

Summary

Through the use of case studies and worked examples, this chapter provides a view of issues faced by organizations and countries in Asia, Europe and America. It begins with a discussion on the emergence and growth of a low carbon economy, detailing the key economic, social and environmental drivers for its development. Several innovative approaches are discussed to enable organizations and countries to assess EMS development. This includes the integration of government strategies and large funding institutions to increase international business opportunities, technological development and minimize environmental risk. The chapter concludes with a review of the growing complexities of environmental management in a low carbon economy and proposes solutions for organizations by way of knowledge networks and diffusion models to determine environmental management best practice.

Low carbon global economy

For many, the journey from a high carbon economy to a low carbon economy will be a long one and, for some, the most achievable objective may just be to move a little at a time further away from the 'high carbon edge' as resources and technology development allows. It is useful at this point to introduce some of the major events from 1992 to 2009 involved in the process of formulating a low carbon economy.

United Nations Framework Convention on Climate Change (UNFCCC), 1992

The international response emerged with the development of UNFCCC. It was adopted at the United Nations Conference on Environment and Development in New York, May 22, 1992. For present and future generations, it aims to stabilize atmospheric concentrations of GHGs at a certain level and prevent the negative impacts of human activities from damaging the climate system (UNDP, 2010).

Kyoto Protocol, 1997

The Kyoto Protocol is a supplementary provision to the UNFCCC. It was established on December 11, 1997 in Kyoto, Japan and put into force on February 16, 2005. It is an international agreement that brings countries together to protect the environment and reduce the effects of global warming and greenhouse emissions. In brief, its goal is to fight global warming. Up to December 2009, 187 countries had signed the Kyoto Protocol, including China but excluding the US and Australia. The provisions of the Kyoto Protocol are legally binding on the ratifying nations, and stronger than those of the UNFCCC (Kyoto Protocol, 2005).

Energy White Paper, 2003 – Our Energy Future: creating a low carbon economy

The British government released the Energy White Paper on February 24, 2003. It is a clear commitment to a 'low carbon economy' that outlines a new economy with low emissions, low energy consumption, low pollution and fully linked with economic growth, social engagement and environmental improvement. It also addresses future energy challenges and gives a new policy direction for energy development and security.

World Environment Day (WED), Kick the habit: Towards a Low Carbon Economy (2008)

On June 5, 2008, the United Nations Environment Programme (UNEP) stated the theme of World Environment Day, 'Kick the habit: Towards a Low Carbon Economy'. Its programme was designed to show a way forward that mitigates climate change, reduces poverty and promotes economic and political stability. The Executive Director of the UNEP said,

> The theme is more than a strong and catchy phrase, if we are to move the global economy to a greener and cleaner one, a sharp reduction in the inefficient use of fossil fuels allied to an increased uptake of renewable energy must be at the centre of the international response – a low-carbon economy is a huge opportunity rather than a burden.
>
> UNFCCC (2009)

The Climate Change Conference in Copenhagen, 2009

The UNFCCC's 15th Conference of Parties (COP 15) was held on December 7, 2009, in Copenhagen, Denmark. Its aim was to formulate a landmark global climate accord to take the place of the Kyoto Protocol. Some countries had prepared their energy consumption and pollution reduction targets during the conference (UNFCCC, 2009).

International comparison of environmental management development

It is difficult to understand, at present, where a country is on its journey to a low carbon economy. This makes it doubly difficult to identify the progress made to date and the number of low carbon or green organizations, the amount of low carbon technology investment and the number of low carbon or green jobs created. A number of attempts have been proposed and the following two help towards an understanding of a possible definition.

- The Ernst & Young (2008) report provides a narrow understanding of green business and defined the Environmental Goods and Services (EGS) industry as including the product areas and processes as detailed in Table 6.1.
- Commissioned by the UK government, BERR (2009) sought to broaden the definition of the EGS sector by including rapidly growing renewable energy technologies as well as a number of other emerging low carbon activities. Table 6.1 details the breakdown of sectors and defines the LCEGS sector.

There is currently no clear or agreed definition of what constitutes a green or low carbon economy. However, the definition of the LCEGS sector has been widened to include existing organizational environmental management activities, encompassing resource efficiency and low carbon solutions. In 2008 the LCEGS sector had an estimated global value of £3 trillion and is estimated to grow by

TABLE 6.1 Breakdown of LCEGS sector

The Low Carbon Environmental Goods and Services subsectors

Environmental	Renewable energy	Emerging low carbon
Air pollution control	Hydro	Alternative fuels
Environmental consultancy	Wave and tidal	Alternative fuels for vehicles
Environmental monitoring	Biomass	Additional energy sources
Marine pollution control	Wind	Carbon capture and storage
Noise and vibration	Geothermal	Carbon finance
Contaminated land remediation	Solar PV	Energy management
Waste management	Renewable consulting	Building technologies
Water supply and waste water treatment		
Recovery and recycling		

Source: BERR (2009)

4 per cent over the next decade (BERR, 2009). Lord Mandelson, the then UK government's Secretary of State for Business, stated that 'low carbon is not a sector in our economy; it is, or will be, our whole economy and a global market.' The challenge for organizations competing in a low carbon economy is not whether to have an EMS but to have a system that integrates new knowledge, skills, finance and flexibility in order to facilitate transition to a low carbon economy. The immediate challenge for many organizations is managing potentially high short-term costs to longer-term benefits from the investments required to make such a transition and to prosper in a low carbon economy. The benefits of such a transition are potentially significant: the Carbon Trust (2008) reports that an organization that is well prepared and proactive could increase its value by 80 per cent.

Benefits and barriers to economic growth in a low carbon economy

The pursuit of a global low carbon economy has many attractions for many nations. The adoption of a global ISO standard linking private and public sector stakeholders and initiatives would offer a new form of low carbon global governance in this new world economy. The adoption of such a standard or set of standards can lead to the overriding of national standards as witnessed with the rise of ISO 14001 and the demise of BS7750. Revised standards lead to new institutional arrangements and the formation of complex networks of public and private stakeholders, with all the potential conflicts that may arise as a consequence of the competing interests of private businesses and civil society stakeholders (Nadvi and Waltring, 2004).

The challenge for organizations in developed countries as well as developing countries is having available capacity to engage with new low carbon standards and make the necessary technological investments. As with the ISO 14001 standard, setting environmental standards also has the effect of setting entry barriers for new entrants in a value chain, and to present new challenges to existing developing country suppliers (Giovannucci and Ponte, 2005). In this context, the role played by the ISO and other similar international organizations is not just that of standards rulemaking but also as an alternative source of authority for global governance and the need to integrate developing country partners into these structures (Summers 2003).

Traditionally, developing countries have lacked the financial and political power to effectively influence the determination of the contents of business standards. Today, most developing nations continue to be 'standard takers' rather than 'standard setters' (Nadvi, 2008) and those with weak standards infrastructure and poorly resourced national standards agencies are likely to be excluded from discussions associated with formulating standards to create low carbon economies.

Environmental management in a low carbon economy

Existing management standards assist with the standardization of a wide range of aspects of business activity, such as quality management, environmental management, the prevention of occupational hazards and the provision of health and safety regulations in the workplace, innovation management and CSR. All of these standards have a similar methodology in relation to their formation, structure, process of implementation and third party monitoring.

To aid clarity, it is useful to distinguish between a management standard and a management system. The former term refers to an organization's commitment to align its operations with a set of accredited criteria specified within, for example, an environment or quality standard, such as ISO 14001, ISO 9001. A management system is an interrelated collection of different elements (methods, procedures, instructions, etc.) by which the organization plans, implements and monitors specific activities related to the objectives that it wishes to attain (Casadesús et al., 2008). A management system is a map, or guide, which explains how the everyday activities of an organization are managed. It is a map which defines the organizational structure of the organization and identifies the key processes and procedures in the organization's operations in the field that the standard refers to (quality management, environmental management, etc.), and which states who takes responsibility for these processes and procedures. Management systems are therefore based on the essential principles of systematization and formalization of tasks, and the importance of environmental modernization of an economy follows the introduction of government policies that connect environmental management to technical environmental innovations and economic performance. Many German organizations that have implemented EMSs such as ISO 14001 and EMAS report an increased number of technical environmental innovations as a result of the introduction of the system (Rennings et al., 2006). These organizations also recognize the importance of knowledge exchange and the learning process in increased environmental innovations and that a carefully designed EMS can reduce environmental and social impact, improve innovation and economic performance.

Consequently, the international norms or standards which establish the guidelines for introducing various management systems into an organization are not, generally speaking, norms which deal with the attainment of an objective or a specific result – in other words, they are not performance standards. Rather, they are standards which establish the need to systematize and formalize through a series of procedures a whole range of organizational processes relating to the various aspects of management. For example, as Jacobsson and Levin (1993) state, a standard of this type relating, say, to safety in the workplace does not deal with the characteristics of the working environment, but rather with the planning and procedures that the organization should put in place so as to deal with the subjects relating to safety. The fact that an organization implements such a standard, and that an independent certifying body conducts an implementation audit and validates it by awarding a certificate, means that the organization concerned has systematized and formalized (in the form of adequate documentation) the activities that the standard concerned is intended to regulate. It is because of this that such standards are often criticized in the field of management studies for their tendency to increase bureaucracy and excessive rigidity.

Regional differences

The international community is currently piloting a number of public policies, new market-based instruments and innovative financial mechanisms, to attract and drive direct investment towards low carbon and climate resilient technologies and practices. In 2012, the private sector invested nearly US$350 billion of new money in clean energy technologies in response to these new policy and financial incentives (OECD, 2012). Although there is some concern that the current financial crisis may freeze financing for green energy projects, or that a number of financial incentives for energy efficiency and renewable energy will be phased out by governments trying to trim budget deficits, it is expected that investment in clean energy technologies will resume its growth to about US$600 billion by 2020.

However, these financial incentives and investments often remain restricted to OECD countries and a small number of rapidly developing countries. Barriers still need to be removed before they can be widely disseminated for easier access by other developing countries. For example, the Kyoto Protocol created the CDM to promote both SD and GHG emission reduction in all developing countries. The CDM is a global cap-and-trade mechanism, which allows developing countries to earn credits for their emission reduction projects and sell these cheaper credits to industrialized countries. Despite its potential, there is strong concern that only a limited number of countries benefit from the CDM, and that this mechanism could bypass Africa entirely. Only five countries – China, India, Brazil, South Korea, and Mexico – are expected to generate over 80 per cent of all CDM credits by 2015 (OECD, 2012). Current finance market rules all too often fail to attract investors into lower-carbon technologies and sustainable land-use projects. The specific market conditions of developing countries need to be incorporated into the design of new market-based and innovative financial mechanisms. A number of reforms to the CDM are currently being discussed to achieve this objective. Simultaneously, developing countries need assistance to put into place an enabling environment (e.g. public policies, institutions, human resources) so that they are better positioned to leverage these new sources of development and capacity building finance.

Europe

Frequent changes of legal regulations, which are important from the viewpoint of the industry sectors, lead to increased risk connected to running business operations. Instability of the regulation system also hinders organizations in the development and implementation of medium- and long-term strategies and increases costs of their operation. There is a need to differentiate between member states' legal framework, public administration culture, skills, constraints and the type of public sectors involved.

EU and Asia

An EU financing eco-innovation strategy needs to have a 'global market' perspective. An 'EU-Asia Financing Clean Technologies Development Partnership' for example could involve EU programmes such as Asia-Invest, the EIB, EU Private Banks and Investment Funds in Asia. Such a financing network linked with organizations committed to low carbon technology development could be replicated in other world regions. Countries, such as Japan, that are more energy efficient help industrializing countries of the region through technical and economic cooperation. The Japan International Cooperation Agency (JICA), the Japan Bank for International Cooperation (JBIC) and the Ministry of Economy, Trade and Industry (METI) are instrumental in sending technical experts and offering low-interest loans for energy efficiency investment. Major recipients of ODA for industry are in low-to-medium income countries.

An 'EU-Asia Financing Clean Technologies Development Partnership' assists business case drivers, technology transfer, early stage development gaps, and organizational structures. The Greenhouse Gas Emission Reduction from Industry in Asia and the Pacific (GERIAP) project is an initiative by UNEP to assist Asian organizations to become more energy- and cost-efficient through strategies that improve energy efficiency, prevent carbon emissions and reduce operational costs. More than 40 organizations from the cement, chemicals, ceramics, steel and paper sectors have participated in this pilot project in Bangladesh, China, India, Indonesia, Mongolia, the Philippines, Sri Lanka, Thailand and Vietnam. By undertaking energy efficient investments and measures, participating organizations have reduced emissions by more than 85,000 tonnes of CO_2 per year, while making annual profits of more than \$4 million (UNEP, 2010a and b).

Multilateral agencies such as the World Bank and the African Development Bank are sponsoring projects on demand side energy management, focusing on electrical power supply and chemical, cement and steel industries. Their current strategy of project-based, government chosen technology transfers, however, to date it is often found to be slow and inflexible. Greater emphasis is required to enable business-to-business cooperation as well as building the capacity of institutions to support technology development initiatives and to make sound policy decisions.

Business case drivers

The business case for most organizations investing in the development of new low carbon technology depends upon demand for the use of that technology in existing industrial processes, or new processes. This prior demand is in turn affected mainly by three factors:

- Price of energy.
- Price of carbon emissions.
- The regulatory framework.

Higher energy prices stimulate demand for close substitute energy sources. However, in many parts of Asia energy prices are often subsidized. In China, electricity is about one-third the cost it is in Europe, due largely to government subsidies and to the difficulty of externalized carbon costs. Such factors serve to undermine the business case for investing in energy efficiency, renewable energy and development of associated technology.

A number of public–private schemes exist to encourage energy efficiency investments in China and India. These are soundly based and have had some early success. However, investment in energy efficiency is still not a high priority for factory owners in, for example, China's southern Guangdong province industrial zone. The US backed P2E2 scheme, and a new IFC scheme, are helping address this.

The price of carbon emissions is also a key factor. Where this is implicitly priced at zero, then investment in both energy efficiency and renewable energy technology is again undermined. Certified Emissions Reductions (CERs) that can be purchased by developed world investors are a way of implicitly pricing carbon emissions in Asian countries where no emissions trading scheme operates. Forward selling such CERs can be a way to help finance investment in environmental technology. In the absence of strong price signals to stimulate investment in, either, energy efficiency or renewable energy, the regulatory framework is an especially important potential stimulus.

In Asia, the role of regulation has traditionally been curtailed, but environmental degradation is forcing change. In China, for example, air and water pollution is increasingly becoming a political issue. Central and provincial government is concerned about growing public unrest and political instability due to the ever more obvious public health effects of high pollution and toxicity levels. Regulatory frameworks are being reviewed and showing some signs of being tightened and improved emission control standards. However, of equal importance to new laws is adequate enforcement of existing laws, which are routinely ignored by many organizations, and tend to be enforced only in areas where foreign organizations are operating. As long as the regulatory framework and its enforcement remains weak, demand for existing low carbon technology, or investment in development of new technology, will also remain weak.

One type of environmental voluntary commitment is the adoption of ISO 14000 standards. Even though they do not have the force of law or government policy, in many cases the ISO 14000 series is becoming the industry code of practice as the market recognizes the value of such voluntary approaches. Since its launch in 1999, the uptake of ISO 14001 has been rapid in Asia and it has become the most commonly used quality assurance standard. Asian corporations comprise approximately 40 per cent of the world's ISO 14000 certified organizations. As of December 2012, Japan leads with 13,104 certificates, followed by China (8,865), the Republic of Korea (2,610), India (1,900), Taiwan (1,463), Thailand (974), Singapore (573), Malaysia (566), Indonesia (369), and the Philippines (312), showing widespread uptake across the region (ISO, 2012). ISO certification is usually awarded to a production facility with a condition that it complies with a set of environmental performance criteria. Energy efficiency requirements are part of the criteria to encourage organizations to engage in a continuous improvement process.

Economic change

To activate a move to a low carbon economy requires a fundamental objective: to reduce carbon emission by reducing the use of traditional fossil fuels. Therefore, it is necessary to pursue investment in the technology to exploit new renewable energy sources as well as improve energy efficiency from fossil fuel energy sources.

In the USA, the Obama administration indicated that it will put $150 billion into alternative energy research over the next 10 years (UNDP, 2009). In Europe, the European Commission indicated in its 'Economic Recovery Plan' that €200 billion will be invested in low carbon technologies and renewable industries to enhance the long-term economic competitiveness of European states (UNDP, 2009).

In China, the National Development and Reform Commission (NDRC, 2007) indicated in its 'medium and long term development plan' that it will invest ¥1.5 trillion over the next 15 years to the following strategic goals of renewable energy development:

- China will increase renewable energy usage in total primary energy consumption to 10 per cent by 2013.
- It will aim to raise this usage to 15 per cent by 2020. This will be achieved by fully utilizing existing mature technology and new, economically feasible renewable energy sources, such as hydropower, biogas, solar thermal, and geothermal, as well as by promoting the development of the wind power, biomass power, and solar PV industries.
- China will also aim to provide electricity to people in remote, off-grid areas and resolve fuel scarcity problems in rural areas through the use of renewable energy sources, doing so according to local conditions thereby protecting the ecological environment. The utilization of organic wastes for energy will be promoted according to the principles of a 'recycling economy', thereby eliminating environmental pollution caused by organic wastes.
- It will actively promote the development of low carbon technologies and industries, creating a renewable energy technology and innovation network. It is forecast that by 2013, China will have achieved the ability to produce in its own country, the main renewable energy generation equipment it needs. By 2020, local manufacturing capability based mainly on home-grown Intellectual Property Right (IPR) will be achieved.
- By 2030, renewable energy will be defined as a mainstream energy source and account for 25 per cent of total energy usage.
- By 2050, renewable energy will be defined as a dominant energy source and account for 40 per cent of total energy usage.

Innovative international approach to environmental management

Under a 'business as usual' high carbon, economic development approach, there are many commentators, practitioners and scientists that predict increased air, water and ground pollution, climate change and global warming. In fact, it is now generally accepted that remaining on this course will inevitably result in negative impacts on human health, and constraints on the improvement of living standards due to increasing prices of essential commodities such as food and energy (IPCC, 2007). Despite this knowledge, the speed of change, however, is imperceptible to the majority of individuals and moves at a rate that would be insufficient to instil significant behaviour change without targeted government intervention.

Since the Stern Review (2006), governments are more aware of the existing global, environmental and economic challenges emanating from a 'business as usual' approach. Many governments have already introduced strategies or implemented policies to stimulate a shift towards a more sustainable society and a low carbon economy by promoting responsible environmental and social business practices, products and services and technological development (WBCSD, 2010).

Greener business practices have important economic pay-offs in terms of resource efficiency and growth. Many of these are in the energy sector or related to energy use. The IEA, for example, estimates that the 17 per cent (US$46 trillion) increase in energy investment required globally between 2010 and 2050 to deliver low-carbon energy systems would yield cumulative fuel savings equal to US$112 trillion (IEA, 2010). Energy conservation is one of the first steps that some organizations have taken to reduce their GHG emissions (OECD, 2010), as it often leads to significant and rapid cost reductions. By being energy efficient and thereby using less energy, for example, Dow Chemicals saved US$9 billion over 15 years and DuPont has saved US$5 billion since 1990.

More generally, organizations seek a competitive edge through clean technology investment, understanding that environmental performance is a major factor for future competitive advantage. Leading organizations are increasingly finding innovative ways of mainstreaming low carbon outcomes into core business activity. An Ernst & Young (2009) survey of 75 per cent of top executives from 300 major global organizations forecast annual clean energy technology spending to rise over the next five years.

New and improved technologies in energy production, such as solar power, biomass, micro-hydro power and biofuels, linked with new approaches to electricity generation and distribution, could reduce the costs and improve the technical feasibility of energy supply in developing countries and allow non-oil and gas producing countries to become more energy self sufficient (IEA, 2008). They would also bring a range of benefits, including reduced dependence on fossil fuels, reduced poverty and lower energy bills for businesses and households.

Environmental action also generates new business opportunities. For instance, management in organizations view the pursuit of improved environmental performance as an opportunity to gain advantage over less technologically advanced rivals and to capture important market share. In natural resource sectors alone, commercial opportunities related to environmental sustainability could be between US$2.1 and US$6.3 trillion by 2050 – assuming that sufficient changes are made to ensure that standards of living can be sustained within the limits of available natural resources and without further harm to biodiversity, climate and other ecosystems (WBCSD, 2010).

Business opportunities have also emerged from the sustainable use of biodiversity and ecosystem services including the global market for certified organic food which exceeds US$30 billion per annum. Valuable new biodiversity related asset classes have also emerged; in the US for example, wetland banking credits range in value from US$7,000–850,000 per hectare and have attracted substantial entrepreneurial investment (TEEB, 2010). There is arguably greater scope for economic growth in this sector.

New business models are also emerging. Energy-saving organizations, for example, provide energy-saving solutions to other organizations and public buildings. These organizations are paid from the savings achieved, not by an up-front payment, facilitating the uptake of costly technologies. Other emerging business models include product service systems where the value proposition shifts more to the services delivered by products rather than the products themselves, such as car sharing schemes (EPA, 2009).

Integrated management systems

Today, the more progressive and visionary organizations have sought to combine quality, health and safety and EMSs into one integrated management system to create one coherent management system. This integrated approach lends itself to realizing consistencies and efficiency of operation and administration. Jonker and Karapetrovic (2004) referred to this approach as an integration of systems rather than an integration of standards.

The interest in adopting integrated management systems has grown in Asia, America and Europe and the focus of such systems is towards low carbon activities and corporate sustainability management (Casadesus *et al.*, 2008).

In assisting organizations to transition to a low carbon economy, integrated management systems adopt a multi-level approach:

- Operational: cross-departmental collaboration and cross-referencing of environmental, quality and health & safety systems by internal work groups. This encourages knowledge exchange and unlocks areas of expertise; it encourages and captures innovation to improve processes and procedures and to develop new products and services.
- Management: generic processes to focus on management tasks within the development of the integrated system. Engage key stakeholders to undertake continuous improvement in integrated system performance and stakeholder involvement with respect to internal and external changes.
- Strategic: creation of a transparent organizational culture, one of integrity, innovation and learning where internal and external networks support the delivery of key, low carbon development themes for the economic benefit of the organization and its community.

Current environmental management practice is for an organization to adopt a set of standard guidelines such as ISO 14001 and implement an EMS to align the organization's business activities to the set guidelines to reduce its exposure to environmental risk. Its supply chain also conforms to the same guidelines set down by the standard. An alternative approach to current practice is for an organization to follow a low carbon theme, or a set of themes such as energy efficiency, waste reduction and low carbon technology, products and service development. All quality, health & safety, environmental management, corporate governance systems act as one integrated system supporting these key low carbon themes. Progress is agreed and monitored by relevant performance or balanced scorecard indicators for each theme. An organization's supply chain also delivers to the same low carbon theme requirements, as opposed to meeting standard guidelines.

The move towards a low carbon economy is visible and dynamic: the constant refreshing and realigning of EMAS and ISO 14001 and the introductions of new CSR, sustainability and energy efficiency standards is testimony to this. An organization producing products and services within a network set around a low carbon theme has greater flexibility and resources to make changes and meet the technological and innovative advances that emerge from the knowledge exchange and collaborative investment activities of the network. These changes can be quickly assimilated into the organization and its network, whereas the constant changing of a standard only leads to an improved standard. Additionally, the constant increase in the number of standards to manage various low carbon organizational aspects serves to create more sets of linear subsystems increasing organizational complexity with diverse and often resource competing sets of improvement objectives. However, one set of objectives positioned around a low carbon theme helps to guide the whole organization, its stakeholders and supply chain to deliver products and services to an agreed set of performance indicators. The introduction of a range of low carbon themes increases the number of organizational objectives but will not add additional competing sets of guidelines or criteria and enables a better fit with the existing set of low carbon performance measures. Therefore, a low carbon theme has an operating framework, systems, procedures and processes that are structured at three levels within the organization. These are, in turn, connected with a network of organizations and stakeholders at the subnational, national and transnational level to deliver the low carbon and economic objectives of the organization and its network.

Technology transfer

In moving towards a low carbon economy, the majority of countries will lead with the development of new technologies as they recognize that rapid development and diffusion of new and existing low carbon technologies both for secure, sustainable energy supplies and the adoption of energy efficiencies, is urgently required. Many of the proposed solutions for a low carbon economy have tended to focus on the type of technology, the technology developer, available funding instruments, intellectual property and enabling support infrastructures that best fits with each country's economic development needs. In the pursuit of a global low carbon economy it is right to say that there is broad agreement that a swift solution is needed and that one country to construct a low carbon economy in isolation is not a solution. However, one country in pursuit of a low carbon economy is reasonable if the objectives for that country are economic growth, technological development, low carbon products and services and a secure, renewable energy source.

The transfer of knowledge and innovation in the economy can be considerably hindered by the lack of cooperation between business entities, as well as other entities involved in transfer of innovations to organizations such as science centres, business environment institutions and central or regional authorities. In European Mediterranean countries a real dynamic is in place as far as renewable energies are concerned. There exists a general lack of consciousness of the advantages of low carbon technologies and the absence of a regulatory framework has been identified as a key barrier to the financial community trying to fund new technologies.

Countries like Japan have a recognized lead in technology and process development in almost all key sectors. However, access to energy efficient technologies constitutes a significant barrier to adoption by Asian industries due to their small size. Generally, Asia has not kept up with technical innovations for energy savings, although China, India and the Republic of Korea have been successful in developing prototype technologies for light industries such as food processing and textiles. Upgrading obsolete technologies in heavy industries like steel, cement, and paper is often found to be expensive, as new technologies need to be transferred from advanced economies (Wang, 2006).

The overall process of low carbon economic development and its diffusion to national stakeholders and consumers is perceived as being high risk in terms of cost of development, and dependent on the success of new products, services and technologies. However, encouraging the transfer of national knowledge and the sharing of development costs and experiences with international or global partners has the potential to reduce the real and perceived risks around the development of new low carbon technologies, building commercial capacity, implementing support infrastructure, engaging consumers and creating a low carbon economy (NEDO, 2010).

Skilled workforce

Very recently, countries such as India, Kingdom of Saudi Arabia (KSA), Brazil and China have embarked on multimillion dollar investment into vocational skills training projects. These countries recognize that the current practice of importing skilled foreign expatriates is only a short-term solution and that a sound skills base and infrastructure is needed for long-term economic sustainability. Governments in KSA and India are building hundreds of 'advance' vocational colleges within their countries for the 'Saudization' and 'Indianization' of its citizens. The lack of technical and vocational education and training of employees is currently a barrier to many countries moving towards a low carbon economy.

The benefits of energy efficiency in environmental and economic terms are sometimes beyond the understanding of managers as well as employees but their involvement is important for implementation at the operational level. Poor understanding of the functional characteristics of energy efficiency measures will increase costs, hamper the achievements of desired results and even disrupt the production process if not implemented correctly. In one survey of employees, a large-scale chemical manufacturer in India discovered that illiteracy was a major hurdle in improving energy performance (Jose, 2005).

Financing low carbon technologies

Some technology options provide substantial energy savings and have a short payback period, but require high initial investment, which is not easily available to many organizations or countries. This is simply due to not having ready access to money as banks generally, do not have the confidence to advance funding to undertake new, perceived high risk projects. Most private financial institutions operate on a risk minimization approach and need collateral backing for loans. Under these circumstances, energy efficiency projects for example do not always produce acceptable appraisal results (UNIDO, 1997). Another financial barrier created by banks and other financial institutions to technology development is 'Capital rationing'. This practice occurs where lenders prefer safer investment alternatives in those areas familiar to the banking sector – low carbon technological development is, currently, a relatively unknown, high risk investment area to the banking sector. In a bid to fill the technology investment gap, emerging financial mechanisms such as energy service organizations (ESCO), provide investment capital for a share of the financial savings.

To be successful, ESCOs need long-term contracts with their clients to cover the initial investment. Until recently, expenditure plans and contracts of government organizations were limited to five years in Japan, forming a critical barrier to adopt ESCO assistance. To remove this obstacle, a recent law allows government organizations to extend the contract period for up to ten years. Accelerating the growth of ESCOs among small business and households also needs new policy approaches. IGES (2007) is currently working on a household ESCO scheme that would attempt to solve the lack of profitability of ESCOs using collaboration and burden sharing by stakeholders. In this scheme, the local bank would serve as the financial service supplier, retail shops would provide electric appliances, environmental specialists serve as energy service advisers to households, and a local public body is the service coordinator. In another study on product service systems which analysed the sustainability potential of ESCOs and their business performance, IGES (2007) suggested an inter-ministry, multi-stakeholder working group to evaluate appropriate financial incentive mechanisms.

Technology transfer

In Asian developing countries a recurrent theme in environmental technology use and development is the problem of technology transfer. A key problem for developers and technology owners is keeping control of their technology. Licensing of technology is an unreliable way of doing this. Some larger corporations that have a strong presence on the ground are electing to invest directly in projects that utilize their own technology. In China, the push for community participation in such projects is strong, and the policy perspective is that communities and organizations should fully collaborate on such projects where possible.

Since widespread deployment of low carbon technologies is crucial to realizing the vision of a low carbon economy in Asia, innovative options should be considered such as:

- Collaboration with developing countries in Asia in the early stages of technology development leading to joint ownership of IPR.
- The creation of a regional technology acquisition fund, which is structured to buy-out IPR and make privately owned technologies available for deployment in Asia's developing countries.
- Establish a regional/international code of compulsory licensing for low carbon technologies along the lines of approaches taken for treatment of human immunodeficiency virus/acquired immunodeficiency syndrome (HIV/AIDS) or the US Clean Air Act. Ensuring additional finance through innovative public and private support mechanisms is critical to make the currently available technologies commercially viable and to provide seed funding to help achieve economies of scale for emerging new technologies.

Also noteworthy is the preference in many cases to use cheaper second generation environmental technology, rather than expensive leading edge technology. This preference seems to exist on both the demand and supply sides. In China the price incentive is not sufficient to justify use of expensive leading edge technology. Owners of such technology may be reluctant to sell it or unable to find sufficient buyers. They are often also reluctant to manufacture it in China due to inadequate manufacturing quality control, which may be less of a problem for cheaper second generation technology.

Asia and early stage technology development

Anecdotal evidence suggests that Asia, including developed countries such as Japan, Singapore and Australia, is not a particularly good place for entrepreneurs or technology developers to access early stage seed-finance. There is some evidence, however, that at the prior stage of primary scientific research and development, Singapore and Australia are leaders. Japan is a surprising laggard in this area, and may in fact be better at technology adaptation and product differentiation than primary scientific research and invention.

Organizational forms

Collaborative initiatives have been developed to assist investment in using or investing in environmental technology development. Innovative public–private sector schemes have been developed including the P2E2 (Project Performance Enhancement Exercise) scheme backed by the US, and run via Hong Kong and Singapore ESCO and investors. This scheme is targeted at Chinese manufacturing and energy utility organizations.

NEDO (2012), a Japanese based environmental programme incubator, has developed a large number of public–private partnerships to provide financing for large commercial energy projects throughout Asia, with a number of participating European and American organizations (Box 6.1).

The use of innovative schemes and public-private partnerships to address the apparent gap in financing early stage research and development has not yet been properly assessed. In this latter area it may be instructive to look also at business–NGO partnerships, particularly where the technology being implemented is relatively simple,

---

Box 6.1   NEDO

The New Industry and Industrial Technology Development Organization (NEDO) has been established to combine the knowledge and resources of industry, government and academia to create a collaborative network to further enhance Japan's economy and industrial competiveness. The organization actively undertakes the development of new energy and energy conservation technologies, the verification of technical results, and the introduction and dissemination of new technologies (e.g. support for introduction). Through these efforts, NEDO plans to promote greater utilization of new energy and improved energy conservation and contribute to a stable energy supply and the resolution of global environmental problems by promoting and demonstrating new energy sources, energy conservation and environmental technologies internationally and domestically based on knowledge obtained from project collaboration.

Seven principles

*   Middle- to long-term view.
*   Cross industrial cooperation.
*   International cooperation.
*   Large scale demonstration.
*   Integrated fields.
*   Common fundamental technology.
*   Standardization.

and lends itself to bottom of the pyramid, microfinance, or social entrepreneurship approaches, using distributed manufacturing or implementation models.

Public–private partnerships

Public–private partnerships have been a highly effective vehicle for the commercialization of low carbon technology at the project level, but successful examples that address long-term, cross-regional issues or priorities on the market level are few.

There is an apparent divergence between societal need and the economic incentives for investment in environmental technology development. At the societal level the emerging climate crisis calls for a significant increase in the level of research and eco-innovation, and the subsequent development of useful technology. However private investor behaviour is shaped by economic drivers that emphasize maximizing financial return and minimizing risk. The result is credit rationing and a lower volume of higher economic value projects. A large majority of potential eco-innovation projects that do not succeed in attracting venture capital or mezzanine investment would still be likely to yield useful technology if they received investment. Publicly backed guarantees for investments in such projects would essentially socialize private risk and underpin an increase in the proportion of projects receiving investment. Such an approach could harness the existing expertise of venture capital investors in prioritizing those projects most likely to succeed.

Sector transition assistance schemes

Joint industry–government initiatives, with public–private funding, that aim to reduce carbon intensity by investing in research and technical development (RTD) targeting the relevant supply chains. RTD aimed at energy efficiency, alternative low-carbon materials development, waste capture and reuse could all be eligible. Public support could be in the form of private–public partnerships, jointly funded cooperative research programmes, or tax based incentives. Industry demand for such transition assistance is largely driven by carbon pricing and industry sector transition assistance could be linked to carbon cap-and-trade schemes.

- Mandating Sustainable Responsible Investment (SRI) Levels for Pension Funds.
- Mandatory legislative requirements for pension funds to invest in SRI a modest proportion (for example, 5 per cent) of their funds under management could significantly increase the supply of RTD investment and the amount of technology being developed and commercialized.

In many cases this may only require statutory confirmation of emerging common-law guidance for pension fund trustees as to their fiduciary duties, and their obligation to ensure long-term low-risk returns for pension holders (SRI is a long-term low-risk asset class). Such legislative measures could be augmented with additional tax incentives for pension funds that meet or exceed a specified SRI target.

Green savings accounts and funds schemes

Financial institutions could be encouraged to offer financial savings products to retail consumers. Funds from such products are invested only in environmentally sustainable investment. Tax incentives could be applied, either directly to the saver, or indirectly to the institution.

Diffusion of environmental management best practice

In a low carbon economy, environmental management best practice centres around three key areas:

- Process innovation: resource and energy efficiency.
- Product innovation: improved or new goods and services.
- Organizational innovation: new forms of management practices.

The diffusion of environmental management best practice accelerates increased innovative activity and reduces external environmental impact costs: in short the whole of society benefits from this approach. It is recognized (Jaffe *et al.*, 2002) that the cost of this approach, if borne by a single organization, can be prohibitive. A collaborative approach through public and private sector funding or a network of organizations would offset these costs and support a financially sustainable, low carbon innovation strategy (Hendricks *et al.*, 2011). The format to this reform is not a 'one size fits all' model. It is an agreement initially, progressing to a formal collaborative international network that is designed to deliver products and services for a group of countries committed to a low carbon economy. This joint initiative or transnational 'inter-linkage' would work towards developing the general rules for a low carbon economy and agree to deliver such products and services that fit with each government's strategies and policies for a low carbon economy. The investment in clean technology and renewable energy innovation would be undertaken by the country best suited to the development of a particular technology, such as Spain for solar energy, Iceland for geothermal energy and Japan for microtechnology. The inter-organizational, transnational collaboration model is a low-cost option for investing in renewable energy sources, energy efficiency and low carbon products and services as it promotes inter-organizational learning and knowledge exchange and investment funding from the private and public sector and formal and informal environmental institutions. At the organizational level, existing EMSs provide an infrastructure for managers to integrate strategies and policies of individual institutions (internal management) and introduce appropriate guidance and procedures of the other transnational inter-institutions into an overarching framework that ensures a systematic, consistent and uniform approach to business activities that meets the needs of the low carbon economic objectives of the network.

Subnational, national and transnational networks

Those growing up in the twenty-first century with PCs, mobile telephones, laptops, tablets and netbooks and other communications media are fully conversant with technological and social networks and their complexity, and know what it means to live in a networked society. It naturally follows therefore that the more complex the economic, environmental or social issue the greater the need for open and meaningful communications and the creation of partnerships or stakeholder knowledge networks to produce a solution.

Examples of subnational, national and transnational networks for environmental improvement, CSR, climate change and SD are already in existence.

Case study: Northern Periphery Programme

The Clim-ATIC project involves a partnership of 13 organizations, and a further ten associated partner organizations, in Scotland, Sweden, Finland, Norway and Greenland. The project involves community stakeholders working in partnership with public sector and academic institutions from each region, to explore the potential for different community sectors to develop knowledge and capacity to adapt to climate change impacts, and deliver real adaptations that provide local economic and social advantages.

The project focused of four interconnected themes:

- Sustainable transport.
- Secure, local and sustainable energy.
- Tourism.
- Risk and response management.

Objectives

The overall objective of the Clim-ATIC project was to establish a sustainable knowledge exchange, information and training service for climate change adaptation network involving countries across the Northern Periphery of Europe. This cooperative network was established and delivered in collaboration with 109 organizations, universities, colleges, and other institutions across the region. The network, and the eventual knowledge exchange, information and training service had a particular emphasis on identifying how exchanging knowledge and technologies around climate adaptation can bring economic opportunities to enhance the sustainability of communities in Northern Periphery countries.

The project received €2.3 million transnational project funding to undertake a number of key activities with organizations, institutions and community sector stakeholders, to build the necessary knowledge and capacity network by means of:

- Implementation of adaptation and demonstration technology with a focus on transnational knowledge exchange.

- Establishment of a formal mechanism to disseminate knowledge for organizational and community benefit.

Information provision

To bring together information and experience from the participating partnership and other Northern Periphery country organizations and to coordinate, where appropriate, the dissemination or signposting of sources of climate change adaptation information and training support for the purposes of:

- Building national and transnational capacity to adapt and deliver local climate change knowledge.
- Providing two-way information exchange with policy makers, businesses and communities.
- Initiate an effective database or repository of climate change information so that end-users are kept informed of activities.
- Present all information in an accessible format.

Dissemination

The objective of the project was to create a subnational, national and transnational network of organizations from Northern Periphery countries to deliver project information through an international conference and the completion of a programme of events and to signpost and disseminate climate change adaptation information and training services to promote understanding and good practice in businesses, communities and local authorities.

Addressing the challenges of climate change adaptation provides opportunities for each country to develop and communicate knowledge at the national level providing insight and support for climate change adaptation challenges in other country contexts. Actions at the national level create patterns of social and ecological impact elsewhere in the world for which each country has a responsibility and from which others can learn.

The information and training service is coordinated at the national and transnational level to address social and economic climate change adaptation issues across the Northern Periphery countries. The provision of these services ensures that all citizens in each country, whatever their age, gender, status, occupation and lifestyle, have access to the best available information and training services. These will provide:

- Delivery by organizations with an existing remit.
- Compilation of the best available information regarding climate change adaptation underpinned by business and community adaptation case studies relevant to Northern Periphery countries.
- Provision of information to businesses, local government agencies and other organizations through the delivery and signposting of information activities.

- Development and delivery of training programmes to facilitate successful climate change adaptation for businesses and community groups and local government.

The aim of this new training and information service is to bring together all regions to collectively contribute to a greater understanding of the impact and adaptation issues associated with climate change with the additional purpose of complementing existing national and transnational training and information services. And, to build capacity to deliver three distinct outcomes:

- Development of a critical mass of information on climate change adaptation issues.
- Creating a gateway and process that draw together existing information and knowledge on climate change adaptation as well as new knowledge generated from other related climate change issues.
- Providing a national and transnational network and a practical context within which a wide range of community members and stakeholders can work with each other to address climate change adaptation issues.

Information services

The key issue is generally not lack of information, but how to make sure that it is accessed and used. People from key organizations in most countries noted interest among their users in having access to, and being able to share, information and best practices from other countries. However, this information has to be of direct relevance to users, and must be in the respective national language(s), and transformed into usable local knowledge.

Often, it is the understanding of processes and methodologies (how to do it) focusing on practical problems, rather than specific issues (what is to be addressed) that is needed. A database of projects and case studies may be of interest for lecturers (to be used for courses etc.), but most users at the local level may need a different way to access information.

It was noted that, as well as science-based information, traditional and historical knowledge based deeply in local experience should also be considered as an input into adaptation information, training and advice services. A further suggestion, with regard to support for the business sector, is that expertise in climate change adaptation should be embedded and available in organizations, institutions and government agencies.

Diffusion

From a global perspective, the successful diffusion of low carbon standards would appear to be closely linked to the basic impetus of realizing a global low carbon economy. This puts the emphasis on the process of globalization of supply chains and the growing importance of transnational partnerships (Braun, 2005b). In the current economic environment, in which outsourcing and the relocation of an organization's activities have become key strategic elements of global supply chains, it is necessary to develop low carbon standards in order to favour the development of low carbon activity. As Boiral and Roy (2007) state, 'the development of management standards is part of the growing globalization of the world economy, which requires the adoption of international standards that facilitates exchange and communication between countries'.

One example of a global institution currently setting international standards and facilitating knowledge exchange and communication is the EU Climate Change Clearinghouse. It is currently being developed and is designed to work with existing information structures. A key challenge for any dissemination network is to link existing databases and national information portals.

Previous discussions have revealed a significant diversity in what is available in different countries. In some countries (Finland, Norway, Sweden, UK/Scotland), a well-developed portal or information system exists or is in an advanced stage of development; in others (Ireland), it is being developed; in others (Greenland, Iceland, Northern Ireland), it is only being talked about (though always regarded as desirable).

Key words are transnational and common issues. Similar issues and problems are being confronted in different regions and countries, and often the most relevant information or experience may be available from another country. Also, changes that are currently occurring to the south may occur further north in the future (Box 6.2)

Box 6.2   OECD – Declaration on green growth

Declaration on Green Growth
Adopted at the Meeting of the Council at Ministerial Level on 25 June 2009
[C/MIN(2009)5/ADD1/FINAL]

WE, THE MINISTERS REPRESENTING THE GOVERNMENTS of Australia, Austria, Belgium, Canada, Chile, the Czech Republic, Denmark, Estonia, Finland, France, Germany, Greece, Hungary, Iceland, Ireland, Israel, Italy, Japan, Korea, Luxembourg, Mexico, the Netherlands, New Zealand, Norway, Poland, Portugal, the Slovak Republic, Slovenia, Spain, Sweden, Switzerland, Turkey, the United Kingdom, the United States and the European Community:

CONSIDERING that:

Economic recovery and environmentally and socially sustainable economic growth are key challenges that all countries are facing today. A number of well targeted policy instruments can be used to encourage green investment in order to simultaneously contribute to economic recovery in the short-term, and help to build the environmentally friendly infrastructure required for a green economy in the long-term, noting that public investment should be consistent with a long-term framework for generating sustainable growth. Green growth will be relevant going beyond the current crisis, addressing urgent challenges including the fight against climate change and environmental degradation, enhancement of energy security, and the creation of new engines for economic growth . . .

Source: OECD (2009)

The creation of a national and transnational network complements existing training and information activities across Northern Periphery countries and helps to build new capacity. It provides a communication structure and creates an evidence base to inform future national climate change adaptation understanding for communities, businesses and local administrations. This is achieved by:

- Supporting community groups and stakeholders to make informed adaptation decisions.
- Creating a structure to enable the exchange of ideas, experience and good practice between participating businesses, communities and local authorities at both national and international levels.

It is anticipated that the provision of these services will build on existing structures and organizations where possible and that the coordinating, central 'hub' is linked to a participating 'node' or 'nodes' in each Northern Periphery country, including those organizations that are not directly involved in the project. Figure 6.1 provides a conceptual framework which demonstrates the operation of a national network that would operate within each Northern Periphery country node.

From the view point of constructing a global low carbon economy, it is important to create networks that link subnational, national and transnational organizations, institutions and agencies.

It is clear to see the complexities of economic, environmental and social demands in a low carbon economy. Such complexities provoke the need for meaningful engagement and commitment from organizations and other public and private sector stakeholders. These linkages can be viewed as networks that provide a wider view of information, ideas, credibility and transparency from sources other than the traditional organizational business needs perspective. This collaboration, cluster or network, therefore, is viewed as the keeper of a set of characteristics that identify low carbon actions and choices and a basis for new low carbon business practices (McCauley and Stephens, 2012).

It has been recognized (Roome, 2001) that the establishment of networks to solve a complex issue such as developing a low carbon economy prepares the ground for innovation as the knowledge residing within participating network

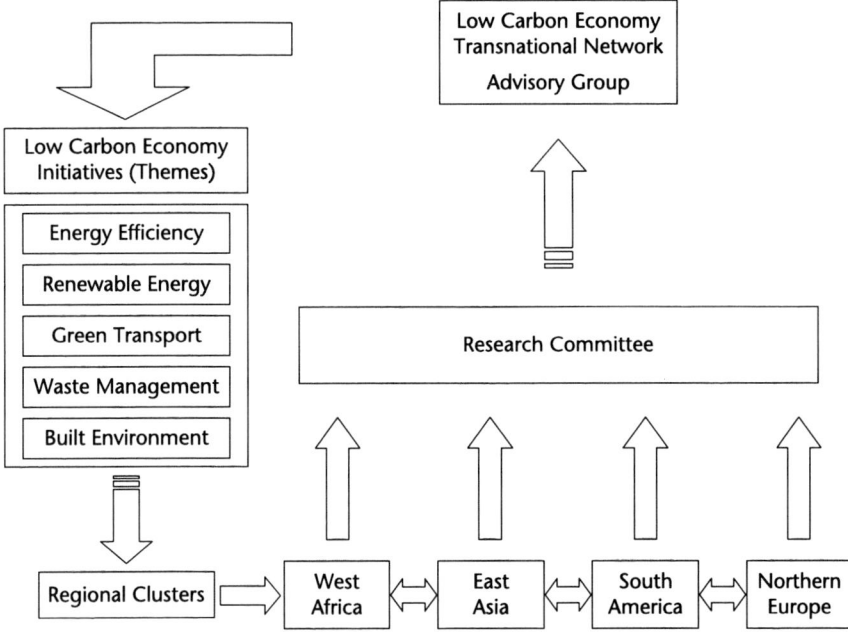

FIGURE 6.1 Low carbon economy transnational network structure

members is exchanged, enhanced and developed. The benefits of network solutions are also predicated on the observation from Trist (1985) that sets of complex issues, such as economic growth, SD and population growth when treated as a set of complex issues under the main (domain) issue such as achieving a low carbon economy, can be treated as an interconnected set of issues and these interconnections defining characteristics. These characteristics mean that any response to address the main issue, such as achieving a low carbon economy, is likely to have an influence on other issues within the set.

Currently, traditional economic, environmental and social issues, debates and solutions are bound up with ambiguity, uncertainty and competing interests. Responses, actions and solutions presented by a network of organizations do not rest with any one organization and emerge through the interaction between a number of organizations with a shared set of values and interests in the solution and a shared commitment to the course of action to be taken.

In a globally connected world of internet, media, finance, technologies, mobility, climatic and ecosystems change issues arising tend to be global in nature. The pursuit of a national or global low carbon economy creates interconnections between the environment, economy, society, culture, finance and technology with the potential for unpredictable outcomes. Traditional organizational regional and national responses to these global issues are lacking the sophistication to provide the desired solutions and therefore create only more uncertainty and unpredictability of actions to be taken. A different response is required, one based on levels of inter-organizational collaboration, innovation and adaptation working within a framework of an agreed set of shared values and guiding principles, on a local and global scale.

Inter-organizational agreements

Against the background of the pursuit of a low carbon economy and the current international economic situation, the trading rules will change. Industries, organizations and economies with high energy consumption and high carbon emissions will be faced with new international trade rules and barriers that will affect their long-term competitiveness. Currently, the main economic power countries pay more attention to carbon emission trading standards and systems. Carbon productivity will become one of the core indicators in international competitiveness. International trade rules will be adjusted under the international climate change framework. Border adjustment tax of carbon emission and carbon tariffs will become the focus. As the standards of ISO 14001 and ISO 19001 are overtaken, development of low carbon economy and high carbon productivity will become the criteria against which organizations and countries will be measured, acquire competitive advantage and avoid future export trade disruption (Oberthur, 2011).

China along with India, Brazil and Mexico are currently going through an urbanization and industrialization development phase. This development, if continued, will lead to these countries becoming high carbon emission countries. These countries are aware that they need to balance current high carbon development with a gradual transition to a low carbon economy. If developed countries begin the push to develop renewable energy industry, low carbon technologies, products and services, there is a possibility that developing countries may end up with a small share of the emerging LCEGS market and receive little benefit for their investment. It is suggested (Wang, 2006) that there is a danger of China, particularly, falling behind in its development of low carbon technology. And, that if this were to happen, China may have to fall back on its traditional skills of product imitation and rely on low-cost competition and miniaturized technology innovation.

The challenge therefore with inter-organizational or multi-organizational collaborations is not to create a perceived exclusive group or membership club. A transnational network of organizations striving to deliver LCEGS to global markets can unintentionally exclude those countries and organizations not able to produce goods and services to these new low carbon guidelines. This has the effect of establishing significant non-tariff barriers to market entry. This issue emerged in an environmental context when the EMAS standard was introduced for organizations trading in European countries. The development of an international environmental standard, ISO 14001 for all countries was as a result of the perceived exclusive membership nature of the EMAS standard.

Having an open membership option for organizations and countries to join national and transnational networks will help to dispel the membership club perception. It will also minimize the risk of breakaway organizations and nations setting up competing networks. However, to work effectively these networks need to be stable yet flexible (Finus *et al.*, 2008), as there is more than one route to a low carbon economy; a one size fits all collaborative network does not suit all members. Different, yet related networks, working towards a global low carbon objective offer organizations and nations different collaborative options or themes to suit their low carbon achievement objectives. According to Wibeck (2012) Sweden's new approach to achieving SD is to develop environmental management based on the establishment of 'guiding visions' regionally, nationally and internationally. This approach is viewed as an effective way to overcome challenges relating to coordinating actors and partners on a common theme.

A transnational network of organizations working within a low carbon economy framework can contribute to capacity building, managing and utilizing information to transition from a high carbon economy to a low carbon economy. Figure 6.1 highlights three broad objectives of such a network. First, the framework is designed around key low carbon objectives and themes to identify the capacity building and reporting process requirements, nationally and regionally and to address specific low carbon theme or themes. Second, the framework identifies the countries and organizations, clusters or subnetworks engaged with each theme and finally, details the flow of information at the subnational, national and transnational level. The third

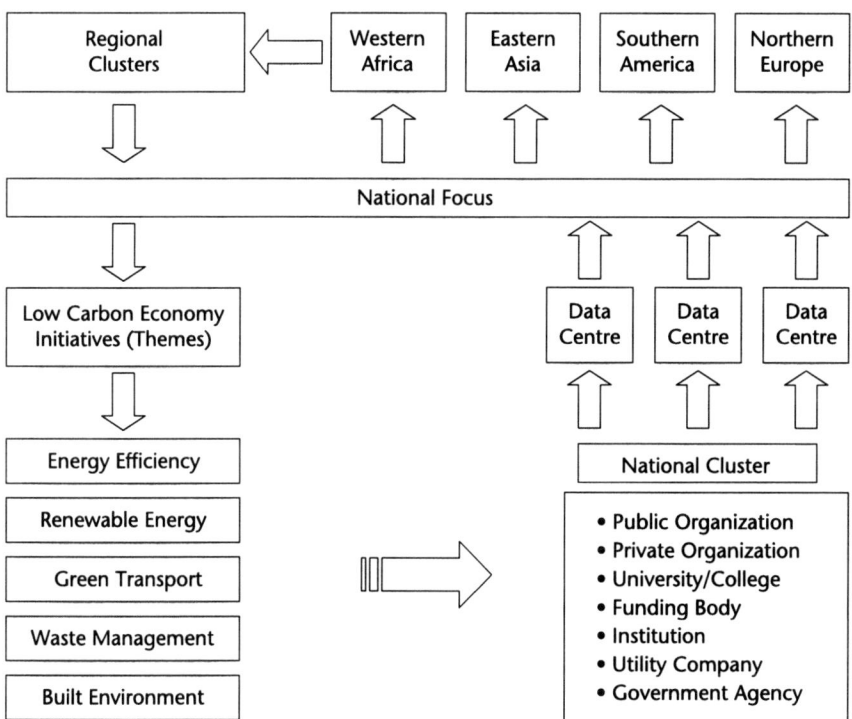

FIGURE 6.2 Low carbon economy national network structure

Box 6.3   Regional initiative – City of Worcester, Central Massachusetts

A regional energy cluster has been formed in the city of Worcester in Central Massachusetts, North Eastern USA with a diverse set of stakeholders including politicians, universities, organizations, and local citizens (Kennedy, 2009). Worcester is the largest city in the region and is the geographic focal point for this regional initiative. The city was originally an industrial city and is involved in an initiative to transform its original 'smokestack' economy to a clean energy using, knowledge-based economy. The main challenge for the City is to stimulate economic restructuring and clean energy system transformation.

The green energy network initiative was launched in 2008 with the goal of building on the existing biotech cluster. The formalizing of the energy network around the biotech initiative has emerged as a centralizing and galvanizing logic for the region.

---

objective relates to the broader demonstration of the development of ideas, exchange of knowledge, pooling of resources and sharing of information and data required to inform the global low carbon economy development.

These networks can be developed along key themes such as renewable energy development, energy efficiency and low carbon technology (Figure 6.2). They can be used to inform and guide framework stakeholders on technology innovation, social impact and economic benefits (Smith *et al.*, 2010). For these networks to emerge and survive and overcome an existing entrenched mainstream regime it needs to combat pressures on three levels (Genus and Coles, 2008). These levels include; the landscape level, the regime level and niche level. The landscape level consists of the political stance, cultural view, socio-economic demands and commitment to the natural environmental. The regime level refers to the existing regime and protection of existing technologies, practices and shared assumptions of the benefits of new technologies. The niche level is where innovative ideas and technologies are supported and allowed to mature and possibly compete with the entrenched regime, that is at the regime level (Geels, 2010) (Box 6.3).

The economic and performance benefits of those organizations that implement EMSs are well evidenced in the literature. These existing systems are organization-centric and serve the interests of the organization in terms of its operating footprint and risk minimization. The environmental standards and systems are voluntary and largely used in a reactive manner. In making the transition to a low carbon economy, successful organizations will require a strategy to simultaneously promote economic vitality and present a transparent commitment to its customers and stakeholders of its contribution to the realization of a low carbon economy.

REFERENCES

Ackroyd, J., Coulter, B., Phillips, P. and Read, A. (2003) 'Business Excellence through Resource Efficiency: An Evaluation of the UK's Highest Recruiting, Facilitated Self-help Waste Minimization Project', *Resources, Conservation and Recycling,* 38: 271–299.

Affisco, J. F., Nasri, F. and Paknejad, M. J. (1997), 'Environmental versus quality standards – an overview and comparison', *International Journal of Quality Science,* 2(1), 5–23.

ADBI (2012), *Policies and Practices for Low Carbon Growth in Asia,* Asia Development Bank Institute (ADBI).

Ammenberg, J. and Sundin, E. (2005), 'Products in Environmental Management Systems: Drivers, Barriers and Experiences', *Journal of Cleaner Production,* 13: 405–415.

Ansoff, H.I. (1991), 'Critique of Henry Mintzberg's The Design School: Reconsidering the Basic Premises of Strategic Management', *Strategic Management Journal,* 12: 449–461.

Argyris, C. (1993), *Knowledge for Action,* Jossey-Bass, San Francisco.

Australian Department of the Environment and Heritage (2005) 'Greenhouse Challenge Plus, Staff Success Story', http://www.greenhouse.gov.au/challenge/tools/staff_success_stories.html.

Avila, J. A. and Whitehead, B. W. (1993), 'What is Environmental Strategy?' *McKinsey Quarterly,* 4: 53–68.

Azzone, G. and Bertele, U. (1994), 'Exploiting Green Strategies for Competitive Advantage', *Long Range Planning,* 27(6): 69–81.

Azzone, G., Bertelè, U. and Noci, G. (1997), 'At Last We are Creating Environmental Strategies Which Work', *Long Range Planning,* 30: 478–571.

Ball, J. (2002), 'Can ISO 14000 and Eco-labelling Turn the Construction Industry Green?' *Building and Environment,* 37(4): 421–428.

Ball, S. and Bell, S. (1997), *Environmental Law,* London, Blackstone Press.

Banerjee, S. B. (2002), 'Corporate Environmentalism: The Construct and its Measurement', *Journal of Business Research,* 55(3): 177–191.

Bansal, (2005), 'Evolving sustainably: a longitudinal study of corporate sustainable development', *Strategic Management Journal,* 26(3): 197–218.

Bansal, P. and Bogner, W.C. (2002), Deciding on ISO 14001: economics, institutions and context, *Long Range Planning,* 35: 269–290.

Bansal, P. and Hunter, T. (2003), Strategic Explanation for the Early Adoption of ISO14001, *Journal of Business Ethics,* 46(3): 289–299.

Bansal, P. and Roth, K. (2000), 'Why Companies Go Green: A Model of Ecological Responsiveness', *The Academy of Management Journal,* 43(4): 717–736.

Bekefi, T. and Epstein, M. J. (2008), 'Transforming Social and Environmental Risks into Opportunities', *Strategic Finance,* 89(9): 42–47.

BERR (2009), *Low Carbon and Environmental Goods and Services: An Industry Analysis,* Department for Business Enterprise and Regulatory Reform, UK Government publication.

Berry, G. R. (2004), 'Environmental Management: The Selling of Corporate Culture', *The Journal of Corporate Citizenship,* 16, 71–84.

Blanco, H. and Bustos, B. (2003), *Standards for sustainable trade: South American report.* International Institute for Sustainable Development, Santiago.

Boesso, G. and Kumar, K. (2007), 'Drivers of corporate voluntary disclosure: A framework and empirical evidence from Italy and the United States', *Accounting, Auditing & Accountability Journal,* 20(2): 269–296.

Boesso, G. and Michelon, G. (2010), 'The effects of stakeholder prioritization on corporate financial performance: an empirical investigation', *International Journal of Management,* 27(3): 470–496.

Boeker, W. (1997), 'Strategic Change: The Influences of Managerial Characteristics and Organizational Growth', *Academy of Management Journal,* 40(1): 152–170.

Boiral, O. and Roy, M. J. (2007), 'ISO 9000: Integration Rationales and Organizational Impacts', *International Journal of Operations and Production Management,* 27(2): 226–247.

Branco, M.C. and Rodrigues, L. L. (2009), 'Exploring the Importance of Social Responsibility Disclosure for Human Resources', *Journal of Human Resource Costing and Accounting,* 13(3): 186–205.

Braun, B. (2005a), 'Building global institutions: the diffusion of management standards in the world economy – an institutional perspective', In C. G. Alvstam and E. W. Schamp (eds), *Linking Industries across the World,* Ashgate, London. 3–27.

Braun, B. (2005b), Environmental issues: writing a more-than-human urban geography, *Progress in Human Geography,* 29: 635–650.

Brundtland Report (1987), *Our Common Future,* World Commission on Environment and Development, Oxford University Press, London.

Brunson, N. and Jacobson, B. (2002), 'Standardization and Uniformity'. In N. Brunson, B. Jacobson and associates (eds.) *A World of Standards,* Oxford: Oxford University Press, 127–137.

BSI (British Standards Institution) (1992), *BS7750 Specification for Environmental Management Systems,* BSI, Milton Keynes.

BSI (1994), *BS 7750, Specification for Environmental Management Systems,* BSI, Milton Keynes, http://shop.bsigroup.com.

BSI (2003), *BS8555 Specification for Environmental Management Systems,* BSI, Milton Keynes.

Buckingham, L. (2003), 'Making a case for Implementing an EMIS', *Chemistry Business,* February 2003.

Buono, A. F. and Kerber, K. W. (2010), 'Creating a Sustainable Approach to Change: Building Organizational Change Capacity', *S. A. M. Advanced Management Journal,* 75(2): 4–21.

Burke, W. W. and Litwin, G. H. (1992), 'A Causal Model of Organisation Performance and Change', *Journal of Management,* 18(3): 523–545.

Business For Social Responsibility Education Fund (2001), 'Suppliers' Perspectives on Greening the Supply Chain - A report on suppliers' views on effective supply chain environmental management strategies', http://www.getf.org/file/toolmanager/O16F15429.pdf.

Business Green (2012), Manchester United scores green goal with environmental standard, *BusinessGreen: Sustainable Thinking,* online newsletter, http://www.businessgreen.com.

Buzzelli, D.T. (1991), 'Time to Structure an Environmental Policy Strategy', *Journal of Business Strategy,* March–April, 17–20.

Carbon Trust (2007), *Investment Trends in European Clean Energy 2003–2006: Watt bubble or carbonated fizz?* Carbon Trust, London.

Carson, R. (1962), *Silent Spring,* Houghton Mifflin, New York.

Casadesus, M., Marimon, F. and Heras, I. (2008), 'ISO 14001 diffusion after the success of the ISO 9001 model', *Journal of Cleaner Production,* 16, 1741–1754.

Cascio, J. (1996), *The ISO 14000 Guide,* McGraw-Hill, New York.

Child, J. A. (1977), *Organization: A Guide to Problems and Practice*, Harper and Row.

Climate Change Act 2008, Great Britain, *Climate Change Act 2008: Elizabeth II.* Chapter 27, London, The Stationery Office.

Cochin, T. J. (1998), 'Continuously Improving Your Environmental Strategies', *Corporate Environmental Strategy*, 5(2): 57–60..

Cohen, W. M. and Levinthal, D. A. (1990), 'Absorptive Capacity: A New Perspective on Learning and Innovation', *Administrative Science Quarterly*, 35: 128–152.

Colman, R. (2005), Corporate Social Responsibility – Where do we really stand? *CMA Management*, February, 7–9.

Cool Earth (2008), *Cool Earth-Innovative Energy Technology Program*, Ministry of Economy, Trade and Industry, UK Government, London.

Corbett, C. J. and Wassenhove, L. N. (1993), 'The Green Fee: Internalising and Operationalising Environmental Issues', *California Management Review*, Fall, 116–135.

Council of the European Communities, (1992), Conservation of natural habitats and of wild fauna and flora, Council Directive 92/43/EEC, http://europa.eu.int/comm/environment/nature/habdir.htm.

Cramer, J. and Zegveld, W. C. L. (1991), 'The Future Role of Technology in Environmental Management', *Futures,* June, 451–468.

Currie, J. (1993), 'An Approach to Assessing the Management of Environmental Responsibilities in Federal Departments and Agencies', *The Journal of Public Sector Management,* September: 69–77.

Darnall, N. (2006), 'Why firms mandate ISO 14001 certification', *Business and Society* 45(3): 354–381.

Dataquest (1999), 1999 Database Market Analysis, Gartner Group.

Demirag, I., Barry, J. and Khadaroo, I. (2005), 'Concluding remarks on emerging governance structures and practices: the state, the market and the voice of civil society', In I. Demirag (ed.) *Corporate Social Responsibility, Accountability and Governance: Global Perspectives*, Greenleaf Publishing, Sheffield, 351–360.

Dodge, H and Welford, R. (1995), 'The ROAST scale', In R. Welford (ed.) *Corporate Environmental Management: Systems and Strategies*, Earthscan, London, 21–22.

Eagean, D. and Streckewald, K. E. (1997), 'Striving to Improve Business Success through Increased Environmental Awareness and Design for the Environment Education Case Study: AMP Incorporated', *Journal of Cleaner Production,* 5(3): 219–223.

Edvardsson, B. (2005), 'Service Quality: Beyond Cognitive Assessment', *Managing Service Quality*, 15(2): 127–131.

Edvardsson, B. and Enquist, B. (2006) 'Quality Improvement in Governmental Services – The role of change pressure exerted by the "market"', *The TQM-Magazine*, 18(1): 7–21.

Edvardsson, B., Enquist, B. and Hay, M. (2006), 'Values Based Service Brands: narratives from IKEA', *Managing Service Quality*, 16(3): 230–246.

Eisenhardt, K. M. (1989), 'Building Theories from Case Study Research', *Academy of Management Review,* 14(4): 532–550.

Elkington, J. and Hailes, J. (1988), *The Green Consumer Guide: High Street Shopping for a Better Environment*, London: Victor Gollancz.

Environmental Index Report (2006), Pilot Environmental Index Report, Yale Centre for Environmental Law & Policy, Yale University, http://www.yale.edu/epi/2006EPI_Report_Full.pdf.

EPA (2007), *Low Carbon Economy Act (Part 1)*, United States, Environment Protection Agency.

EPA (2009) *American Recovery and Reinvestment Act of 2009 (Recovery Act)*, United States Environmental Protection Agency.

Ernst and Young (2008), *Comparative Advantage and Green Business,* Ernst and Young report URN 08/1036, Department of Business, Enterprise and Regulatory Reform (BERR).

Ernst and Young (2009), Climate Change Investment on the Rise, Action Amid Uncertainty, http://www.ey.com/GL/en/Services/Specialty-Services/Climate-Change-and-Sustainability-Services/Action-amid-uncertainty—Climate-change-investment-on-the-rise.

European Council (2007), *An Energy Policy for Europe* (SEC(2007) 12), Communication from the Commission to the European Council and the European Parliament, Brussels.

European Union (2009), *EU Emissions Trading System,* Directive 2009/29/EC of the European Parliament and the Council, Brussels.

Finus, M., Saiz, M. E., and Hendrix, E. M. T. (2008), 'An Empirical Test of New Developments in Coalition Theory for the Design of International Environmental Agreements', *Environment and Development Economics,* 14, 117–137.

Floyd, S. W. and Wooldridge, B. (1992), 'Managing Strategic Consensus: The Foundation of Effective Implementation', *Academy of Management Executive,* 6(4): 27–39.

Foster, C. and Green, K. (2000), 'Greening the Innovation Process', *Business Strategy and the Environment,* 9(5): 287–303.

Freeman, C. and Soete, L. (1997), *The Economics of Industrial Innovation.* Pinter: London.

Freeman, R. E. (1994), The politics of stakeholder theory, *Business Ethics Quarterly,* 4(4): 409–421.

Fuller, M. K. and Swanson, E. B. (1992), 'Information Centres as Organizational Innovation: Exploring the Correlates of Implementation Success', *Journal of Management Information Systems,* 9(1): 47–67.

Gallagher, D. R., Andrews, R. N. L., Chandrachai, A. and Rohitratana, K. (2004), Environmental management systems in the US and Thailand: a case comparison, *Greener Management International,* 46: 41–56.

Gallarotti, G. M. (1995), 'It Pays to be Green', *Columbia Journal of World Business,* Winter: 38–57.

Gascoigne, J. (2002), 'Supply Chain Management – Project Acorn', *Corporate Environmental Strategy,* 9(1): 62–68.

Geels, F. W. (2010), 'Ontologies, Socio-technical Transitions (to Sustainable Development) and the Multi-level Perspective', *Research Policy* 39, 495–510.

GEMI (2002), Exploring Pathways to a Sustainable Enterprise, SD Planner User Guide (Global Environmental Management Initiative [GEMI]; www.gemi.org/docs/PubTools. htm).

Genus, A., and Coles, A. M. (2008), 'Rethinking the Multi-level Perspective of Technological Transition', *Research Policy,* 37(9): 1436–1445.

Ghobadian, A. H., Viney, J. L., and James, P. (1998) 'Extending Linear Approaches to Mapping Corporate Environmental Behaviour', *Business Strategy and the Environment* 7(1): 13–23.

Gilding, P., Hogarth, M. and Humphries, R. (2002), 'Safe companies: An alternative approach to operationalizing sustainability'. *Corporate Environmental Strategy,* 9(4): 390–397.

Giovannucci, D. and Ponte, S. (2005), 'Standards as a New Form of Social Contract? Sustainability Initiatives in the Coffee Industry', *Food Policy,* 30(3): 284–301.

González-Benito, J. and González-Benito, O. (2005), 'A Study of the Motivation for the Environmental Transformation of Organization', *Industrial Marketing Management,* 34: 462–475.

Governors' Global Climate Summit (2008), Global Climate Solutions Declaration, November, Los Angeles, http://www.climatechange.ca.gov/events/2008_summit/ DECLARATION.PDF.

Governors' Global Climate Summit 3, (2010), Building the Green Economy, November 15–16, 2010, University of California, Davis, http://rona.unep.org/events/.

Greeno, J. L. (1991), '*Environmental Excellence: Meeting the Challenge'*, Arthur D. Little Prism, Third Quarter, 13–31.

Greeno, J. L. and Robinson, S. N. (1992), 'Rethinking Corporate Environmental Management', *Columbia Journal of World Business,* Fall and Winter: 223–232.

Greiner, L. E. (1972), 'Evolution and Revolution as Organizations Grow', *Harvard Business Review,* July–August: 37–46.

Griffith, A. and Bhutto, K. (2008), 'Improving Management Performance through Integrated Management Systems (IMS) in the UK', *Management of Environmental Quality,* 19: 565–578.

Hale, M. (1995), 'Training for Environmental Technologies and Environmental Management', *Journal of Cleaner Production,* 3(1–2): 19–23.

Hart, S. L. (1997), 'Beyond Greening: Strategies for a Sustainable World', *Harvard Business Review,* January–February, 66–76.

Haufler, A. (1999), 'Prospects for co-ordination of corporate taxation and the taxation of interest income in the EU', *Fiscal Studies, Institute for Fiscal Studies,* 20(2): 133–153.

Haveman, M. and Dorfman, M. (1999), 'Breaking Down the "Green Wall" (Part One): Early Efforts at Integrating Business and Environment at SC Johnson', *Corporate Environmental Strategy,* 6(1): 5–13.

Helpman, E. (1998), 'General Purpose Technologies and Economic Growth: Introduction'. In E. Helpman, (ed.) *General Purpose Technologies and Economic Growth,* Cambridge, MIT Press, 1–13.

Hendricks, B., Pool, S., and Kaufman, L. (2011), *Low-carbon Innovation: A Uniquely American Strategy for Industrial Renewal,* Global Climate Network.

Heras, I., Dick, G. P. M. and Casadesus, M. (2002), 'ISO 9000 registration's impact on sales and profitability: A longitudinal analysis of performance before and after accreditation', *International Journal of Quality and Reliability Management, 19*(6): 774–791.

Herbig, P.A. (1994), *The Innovation Matrix: Culture and Structure Prerequisites to Innovation.* Quorum Books, Westport, Connecticut.

Hertin, J., Berkhout, F., Wagner, M., and Tyteca, D. (2008), 'Are EMS Environmentally Effective? The Link Between Environmental Management Systems and Environmental Performance in European Companies', *Journal of Environmental Planning and Management,* 51(2): 255–280.

Hillary, R. (2004), 'Environmental Management Systems and the Smaller Enterprise', *Journal of Cleaner Production,* 12: 561–569.

Holtom, G. (2010), 'The Need for an Environmental Management System and what this Means for Mines', *Engineering and Mining Journal,* 211(3), April, 46–49.

Hui, I. K., Chan, A. H. S., and Pun, K. F. (2001), 'A Study of the Environmental Management System Implementation Practices', *Journal of Cleaner Production,* 9: 269–276.

Hunt, C. B. and Auster, E. A. (1990), 'Proactive Environmental Management: Avoiding the Toxic Trap', *Sloan Management Review,* Winter: 7–18.

Hutchinson, C. (1996), 'Integrating Environmental Policy with Business Strategy', *Long Range Planning,* 9(1): 11–23.

Icelandic New Energy (2008), *Promoting hydrogen in Iceland.* http://www.newenergy.is/en.

IEA (2008), *World Energy Outlook 2008,* International Environment Agency publication.

IEA (2010), *World Energy Outlook,* International Environment Agency Publication.

IGES (2007), IGES open forum: *Global environmental strategies for Asia Pacific, What new from IGES?* October, Institute of Global Environmental Strategies.

IGES (2011), *Climate Change Policies in the Asia-Pacific: Re-uniting Climate Change and Sustainable Development,* IGES White Paper, Japan.

Irwin, A., Georg, S., and Vergragt, P. (1994), 'The Social Management of Environmental Change', *Futures,* 26(3): 323–334.

ISO (2001), Transitioning to ISO 2001, International Standards Organization, http://www.iso.org/iso/home/standards.htm.

ISO (2012), The ISO Survey of Management System Standard Certifications, http://www.iso.org/iso/home/standards/certification/iso-survey.htm?certificate=ISO%209001&countrycode=AF.

Jabbour, C. J. C., Santos, F. C. A., and Nagano, M. S. (2008), 'Environmental Management System and Human Resource Practices: Is there a link between them in four Brazilian Companies?' *Journal of Cleaner Production,* 16(17): 1922–1925.

Jacobs, B. W., Singhal, V. R., and Subramanian, R. (2010), 'An Empirical Investigation of Environmental Performance and the Market Value of the Firm', *Journal of Operations Management,* 28(5): 430–441.

Jacobsson, M. J. and Levin, J. A. (1993), 'Conceptual Frameworks for Network Learning Environments: Constructing Personal and Shared Knowledge Spaces', *International Journal of Educational Telecommunications,* 1(4): 367–388.

Jaffe, A. B., Newell, R. G. and Stavins, R. N. (2002) Environmental Policy and Technological Change, The Fondazione Eni Enrico Mattei Note di Lavoro Series Index: http://www.feem.it/web/activ/_activ.html.

Jiang, R. J. and Bansal, P. (2003) Seeing the need for ISO 14001, *The Journal of Management Studies,* 40(4): 1047–1061.

Jonker, J. and Karapetrovic, S. (2004), 'Systems Thinking for the Integration of Management Systems', *Business Process Management Journal,* 10(3): 608–615.

Jose, M. B. H. (2005), 'Technological Innovation in Chile: Where We Are What Can Be Done', *Journal Economia Chilena (The Chilean Economy),* Central Bank of Chile, 8(1): 53–77.

Kantz, O. (2000), 'Volvo's Holistic Approach to Environmental Strategy', *Corporate Environmental Strategy,* 7: 156–169.

Kennedy, E. H.; Beckley, T. M.; McFarlane, B. L. and Nadeau, S. (2009), 'Why We Don't "Walk the Talk": Understanding the Environmental Values/Behaviour Gap in Canada', *Human Ecology Review,* 16(2): 151–160.

Khanna, M. and Anton W. R. Q. (2002), 'What is Driving Corporate Environmentalism: Opportunity or Threat?' *Corporate Environmental Strategy,* 9(4): 409–417.

Khanna, M. and Kumar, S. (2011), 'Corporate Environmental Management and Environmental Efficiency', *Environmental Resource Economics,* 50: 227–242.

Ki-Hoon, L. and Ball, R. (2003), 'Achieving Sustainable Corporate Competiveness: Strategic Link between Top Management's (Green) Commitment and Corporate Environmental Strategy', *Greener Management International,* Winter: 89–104.

Kim, M-K., Park, M.-C. and, Jeong, D.-H. (2004), 'The effects of customer satisfaction and switching barrier on customer loyalty in Korean mobile telecommunication services', *Telecommunications Policy,* 28: 145–159.

Kirkland, L. H. and Thompson, D. (1999), 'Challenges in Designing, Implementing and Operating an Environmental Management System', *Business Strategy and the Environment,* 8: 128–143.

Kline, S. J. and Rosenberg, N. (1986), 'An overview of innovation'. In R. Landau and N. R. (eds). *The Positive Sum Strategy: Harnessing Technology for Economic Growth.* Washington, DC, National Academy Press: 275–305.

Kolk, A. and Mauser, A. (2002), 'The Evolution of Environmental Management: From Stage Models to Performance Evaluation', *Business Strategy and the Environment,* 11: 14–31.

Kyoto Protocol (2005), Kyoto Protocol to the United Nations Framework Convention on Climate Change (1998), United Nations.

Länsiluoto, A. and Järvenpää, M. (2008), 'Environmental and performance management forces: Integrating "greenness" into balanced scorecard', *Qualitative Research in Accounting & Management,* 5(3): 184–206.

Laszlo, C. (2003), *The Sustainable Company: How to create lasting value through Social and Environmental Performance,* Island Press, Washington.

Lee, B. W. and Green, K. (1994), 'Towards Commercial and Environmental Excellence: A Green Portfolio Matrix', *Business Strategy and the Environment,* 3(3): 1–9.

Leon-Soriano, R., Munoz-Torres, M. J., and Chalmeta-Rosalen, R. (2010), 'Methodology for Sustainability, Strategic Planning and Management', *Industrial Management and Data Systems,* 110(2): 249–268.

Linnenluecke, M. K. and Griffiths, A. (2010), 'Corporate Sustainability and Organizational Culture', *Journal of World Business,* 45: 357–366.

Lorsch, J. and Allen, S. (1973), *Managing Diversity and Independence,* Cambridge, MA, Harvard University Press.

Lozano, M. and Valles, J. (2007), 'An analysis of the implementation of an environmental management system in a local public administration', *Journal of Environmental Management,* 82(4): 495–511.

LRQA (2004), *How can EMAS benefit my organization,* Lloyds Register Quality Assurance, http://www.lrqa.co.uk.

Malmi, T. (1999), 'Activity-based costing diffusion across organizations: an exploratory empirical analysis of Finnish organizations', *Accounting Organisations and Society,* 24: 649–672.

McCauley, S. M. and Stephens, J. C. (2012), 'Green Energy Clusters and Socio-technical Transitions: Analysis of a Sustainable Energy Cluster for Regional Economic Development in Central Massachusetts, USA', *Sustain Science*, 7, 213–225.

McGrew, A. (1990), 'The Political Dynamics of the New Environmentalism', *Industrial Crisis Quarterly*, 4: 292–305.

Melnyk, S. A., Sroufe, R. P., and Calantone, R. (2003) 'Assessing the Impact of Environmental Management Systems on Corporate and Environmental Performance', *Journal of Operations Management* 21(3): 329–51.

Miles, R. E. and Snow, C. C. (1978), *Organization, Strategy and Structure*, New York, McGraw-Hill.

Miller, D. and Freisen, P. H. (1980), 'Momentum and Revolution in Organizational Adaptation', *Academy of Management Journal*, 23: 591–614.

Mintzberg, H. (1987), 'Crafting Strategy', *Harvard Business Review*, July-August: 38–47.

Morrow, D. and Rondinelli, D. (2002), 'Adopting Corporate Environmental Management Systems: Motivations and Results of ISO 14001 and EMAS Certification', *European Management Journal*, 20(2): 159–171.

Müller, K. and Koechlin, D. (1992), 'Environmentally Conscious Management', in D. Koechlin and K. Müller (eds.), *Green Business Opportunities: The Profit Potential*, Pitman, London, 166–172.

Nadvi, K. (2008), 'Global Standards, Global Governance and the Organisation of Global Value Chains', *Journal of Economic Geography* 8(3): 323–343.

Nadvi, K. M. and Waltring, F. (2004), 'Making Sense of Global Standards', In H Schmitz (ed.) *Local Enterprises in the Global Economy: Issues of Governance and Upgrading*, Edward Elgar, 53–94.

NDRC (2007), China's National Climate Change Programme, prepared under the auspices of National Development and Reform Commission, Peoples Republic of China http://en.ndrc.gov.cn/newsrelease/P020070604561191006823.pdf.

NEDO (2010), *Global Development (International Project Activities) Promotion of International Collaboration*, New Energy and Industrial Technology Development Organization (NEDO), http://www.nedo.go.jp/english/introducing_project2_3.html.

NEDO (2012), *Smart Community Summit 2012 and Smart Grid Exhibition 2012, Event Report*, New Energy and Industrial Technology Development Organization (NEDO), http://www.nedo.go.jp/english/event_20120530.html.

NIES (2010), Japan's National Greenhouse Gas Emissions in Fiscal Year 2010 (Final Figures), National Institute of Environmental Studies, http://www.nies.go.jp/.

Nishitani, K. (2011), 'An Empirical Analysis of the Effects on Firms' Economic Performance of Implementing Environmental Management Systems', *Environmental Resource Economics*, 48: 569–586.

Nuki, T. (2007), 'Environmental Issues and Theory of Management', *Asian Business & Management*, 6(2): 123–142.

Oberthur, S. (2011), 'The European Union's Performance in the International Climate Change Regime', *Journal of European Integration*, 33(6): 667–682.

OECD (2005), *Oslo Manual*, Guidelines for collecting and interpreting innovation data, 3rd Edition, OECD and Eurostat.

OECD (2010), *Sustainable Manufacturing and Eco-Innovation: Framework, Practices and Measurement—Synthesis Report*, OECD, Paris.

OECD (2012), *Towards a Positive Legacy of a Terrible Crisis*, OECD Yearbook 2012, OECD Publishing.

Orij, R. (2010), 'Corporate social disclosures in the context of national cultures and stakeholder theory', *Accounting, Auditing & Accountability Journal*, 23(7): 868–889.

Pan, J.N. (2003), 'A comparative study on motivation for and experience with ISO 9000 and ISO 1400 certification among Far Eastern countries', *Industrial Management & Data Systems*, 103(8): 564–578.

Paraschiv, D. M., Nemoianu, E. L., Langa, C. A., and Szabo, T. (2012), 'Eco-Innovation, Responsible Leadership and Organizational Change for Corporate Sustainability', *Amfiteatru Economic*, 14, 32: 404–419.

Pascual, B. and Gomez-Mejia, L. R. (2009), 'Environmental Performance and Executive Compensation: An Integrated Agency-Institutional Perspective'. *Academy of Management Journal*, 52(1): 47–58.

Patel, P. and Younger, M. (1978), 'A Frame of Reference for Strategy Development', *Long Range Planning,* 11: 6–12.

Peattie, K. J. (1990), 'Painting Marketing Education (or How to Recycle Old Ideas)', *Journal of Marketing Management,* 6(2): 105–125.

Peters, T. J. and Waterman, R. H. (1989), *In Search of Excellence: Lesson's from America's Best Run Organization,* New York, Harper and Row.

Petulla, J. M. (1987), *Environmental Protection in the United States,* San Francisco: San Francisco Study Center.

Pfeffer, J. (1996), 'When it Comes to "Best Practices"—Why do Smart Organizations Occasionally do Dumb Things?' *Organizational Dynamics,* Summer, 33–43.

Piercy, N. (1989), 'Diagnosing and Solving Implementation Problems in Strategic Planning', *Journal of General Management,* 15(1): 19–38.

Poksinska, B., Dahlgaard, J. J. and Eklund, J. A. E. (2003), 'Implementing ISO 14000 in Sweden: motives, benefits and comparisons with ISO 9000', *The International Journal of Quality & Reliability Management,* 20: 585–597.

Porritt, J. (2005), *Capitalism as if the World Matters,* Earthscan, London.

Porter, M. E. and Kramer, M. R. (2011), 'Creating Shared value: How to Reinvent capitalism and unleash a wave of innovation and growth', January–February, *Harvard Business Review.*

Porter, M. E.; van der Linde, C. (1995), Toward a New Conception of the Environment-Competitiveness Relationship, *The Journal of Economic Perspectives,* 97–118.

PricewaterhouseCoopers LLP (2002), *2002 Sustainability Survey* Report, Pricewaterhouse Coopers LLP; www.pwc.com/fr/pwc_pdf/pwc_sustainability.pdf.

Pruzan, P. (1998), 'From Control to Value Based Management and Accountability', *Journal of Business Ethics,* 17: 1379–1394.

Pujari, D. and Wright, G. (1994), *Strategic Green Marketing: An Integrated Approach,* Marketing Education Group, Annual Conference, University of Ulster.

Quinn, J. B. (1978), 'Strategic Change: Logical Incrementalism', *Sloan Management Review,* Fall: 7–19.

Quinn, J. B. (1980), 'Managing Strategic Change', *Sloan Management Review,* Summer: 3–20.

Ralston, D., Egri, C., Stewart, S., Terpstra, R. and Kaicheng, Y. (1999), 'Doing Business in the 21st Century with the New Generation of Chinese Managers: A Study of Generational Shifts in Work Values in China', *Journal of International Business Studies,* 30(2): 415–427.

Ramus, C. A. (1998), 'How Environmental Communication Supports Employee Participation: A Case Study of EMI Music', *Corporate Environmental Strategy,* 5(4): 69–74.

Ramus, C. A. (2002), 'Encouraging Innovative Environmental Actions: What Organizations and Managers Must Do', *Journal of World Business,* 37: 151–164.

Reed, R. and Buckley, M. R. (1988), 'Strategy in Action – Techniques for Implementing Strategy', *Long Range Planning,* 2(3): 67–74.

Rennings, K., Ziegler, A., Ankele, K., and Hoffman, E. (2006), 'The Influence of Different Characteristics of the EU Environmental Management and Auditing Scheme on Technical Environmental Innovations and Economic Performance', *Ecological Economics,* 7(1): 45–59.

Renwick, D., Redman, T., and Maquire, S. (2008). Green HRM: A review, process model, and research agenda, *Discussion Paper Series, University of Sheffield Management School,* The University of Sheffield.

Rintanen, S. (2005), 'The establishment and development directions of corporate environmental management. Case studies in Finnish and Italian meat processing sector', Academic dissertation, Turku School of Economic and Business Administration.

Rittenhouse, D.G. (2003), Piecing together a Sustainable Development Strategy, Centre for Economic Performance (CEP), March 2003, 32–38.

Rondinelli, D.A. and Vastag, G.A. (1996), 'International environmental management standards and corporate policies: An integrative framework', *California Management Review,* 39(1): 106–122.

Roome, N. (1992), 'Developing Environmental Management Strategies', *Business Strategy and the Environment,* 1(1): 11–24.

Roome, N. (1994), 'Business Strategy, R & D Management and Environmental Imperatives', *R & D Management,* 24: 65–82.

Roome, N. (2001), 'Conceptualizing and Studying the Contribution of Networks in Environmental Management and Sustainable Development', *Business Strategy and the Environment,* 10, 69–76.

Rothenberg, S. (2007), 'Sustainability through Servicing', *Sloan Management Review,* 48(2): 83–91.

Rothenberg, S., Maxwell, J., and Marcus, A. (1992), 'Issues in the Implementation of Proactive Environmental Strategies', *Business Strategy and the Environment,* 1(4): 1–12.

Schaltegger, S.; Burritt, R. and Petersen, H. (2003), *An introduction to corporate environmental management—Striving for sustainability,* Sheffield: Greenleaf.

Schein, E. H. (1984), 'Coming to a New Awareness of Organizational Culture', *Sloan Management Review',* 25(2): 3–16.

Schmidheiny, S. (1990), *The Entrepreneurial Mission in the Quest for Sustainable Development: The Greening of Enterprise,* International Chamber of Commerce, Paris.

Schneider, B. and White, S. (2004), *Service Quality: Research Perspectives,* Sage Publications, California.

Schumpeter, J. A. (1942), *Capitalism, Socialism and Democracy.* New York: Harper and Row.

Segerson, K. and Miceli, T. J. (1998), Voluntary Environmental Agreements: Good or Bad News for Environmental Protection, *Journal of Environmental Economics and Management,* 36(2): 109–130.

Sheldon, C. (2003), *The Acorn Trust and BS8555,* http://www.groundwork.org.uk.

Shelton, R. D. (1994), 'Hitting the Green Wall: Why Corporate Programmes Get Stalled', *Corporate Environmental Strategy,* 2(2): 5–11.

Shin, S. (2005), 'The role of the government in voluntary environmental protection schemes: The case of ISO 14001 in China', *Issues & Studies,* 41(4): 141–173.

Smith, A., Voss, P., and Grin, J. (2010), 'Innovation Studies and Sustainability Transitions: The Allure of the Multi-level Perspective and its Challenges', *Research Policy,* 39: 435–488.

Smith, G. (1990), 'How Green is my Valley', *Marketing and Research Today,* June: 76–82.

Smith, P. A. C. and Sharicz, C. (2011), 'The Shift Needed for Sustainability', *The Learning Organization,* 18(1): 73–86.

Sprinkle, G. B. and Maines, L. A. (2010), 'The benefits and costs of corporate social responsibility', *Business Horizons,* 53: 445–453.

Sroufe, R., Curkovic, S., Montabon, F. and Melnyk, S. (2000), 'The New Product Design Process and Design for Environment: Crossing the Chasm', *International Journal of Operations and Production Management,* 20(2): 267–291.

Steger, U. (2000), 'Environmental Management Systems: Empirical Evidence and Further Perspectives', *European Management Journal,* 18(1): 23–37.

Stern, N. (2006), *The Economics of Climate Change (The Stern Review),* Cambridge University Press, Cambridge.

Stone, L. (2000), 'When Case Studies are not Enough: the Influence of Corporate Culture and Employee Attitudes on the Success of Cleaner Production Initiatives', *Journal of Cleaner Production,* 8: 353–359.

Summers, L. H. (2003), 'Godkin Lectures', John F. Kennedy School of Government, Harvard University, April, Harvard University Press.

Swann, P. (2009), New Media, New Tools, New Audience, *Journalism and Mass Communication Quarterly,* 86(2): 470–472.

Tapon, F. and Sarabura, M. (1995), 'The Greening of Corporate Strategy in the Chemical Industry: Two Steps Forward One Step Back', *Journal of Strategic Change,* 4: 307–321.

Taylor, S. R. (1992), 'Green Management: The Next Competitive Weapon', *Futures,* September: 669–680.

TEEB (2010), *Mainstreaming the economics of nature, the economics of ecosystems and biodiversity: A synthesis of the approach, conclusions and recommendations of TEEB,* United Nations Environment Programme.

The Acorn Trust (2005), 'Phased Implementation – How does it work?' http://www.theacorntrust.org/in_method_intro.shtml

The Commission on Oil Dependence (2006), *Making Sweden an Oil Free Society,* Swedish Government Publication.

Timberlake, L. (2002), *The Business Case for Sustainable Development,* World Business Council for Sustainable Development.

Tinsley, S. (2002), 'EMS Models for Business Strategy Development', *Business Strategy and the Environment,* 11(6): 376–390.

Tinsley, S. and Melton, K. (1997), 'Sustainable Development and its Effect on the Marketing Planning Process', *Eco-Management and Auditing Journal,* 4(3): 116–126.

Tinsley, S. and Pillai, I., (2006), *Environmental Management Systems: Understanding Organizational Barriers and Drivers,* Earthscan UK.

Trist, E. (1985), 'Intervention Strategies for Inter-organizational Domains', *Human Relations,* 36(3): 269–284.

Tsai, S. H. T. and Child, J. (1997), 'Strategic Responses of Multinational Corporations to Environmental Demands', *Journal of General Management,* 23(1): 1–21.

United Nations Conference on Environment & Development (1992), United Nations Sustainable Development, Agenda 21, Rio de Janerio, Brazil, 3–14 June 1992.

UNDP (2009), *Charting a New Low-Carbon Route to Development: A Primer on Integrated Climate Change Planning for Regional Governments,* United Nations Development Programme.

UNDP (2010), China Human Development Report 2009/10: *China and a Sustainable Future: Towards a Low Carbon Economy and Society,* United Nations Development Program China Translation and Publishing Corporation.

UNEP (1995), 'Environmental Policy and Industrial Innovation: Strategies in Europe, the US and Japan', *UNEP Industry and Environment,* April–September: 118.

UNEP (2010a), *Governors' Global Climate Summit with focus on green economy,* United Nations Environment Programme, Washington DC, 10 November 2010.

UNEP (2010b), *Annual Report,* United Nations Environment Programme Publications.

UNFCCC (2008), Poznan Climate Change Conference, December, United Nations Framework Convention on Climate Change, Report, FCCC/CP/2008/7/Add.1.

UNFCCC (2009), Conference of the Parties on its fifteenth session, December, Report FCCC/CP/2009/11, United Nations Framework Convention on Climate Change, Copenhagen.

UNIDO (1997), *Annual Report 1997,* United Nations Industrial Development Organization publication.

Vargo, S. L. and Lusch, R. F. (2004). 'Evolving to a New Dominant Logic for Marketing', *Journal of Marketing,* 68, 1–17.

Vastag, G., Kerekes, S., and Rondinelli, D. (1996), 'Evaluation of Corporate Environmental Management Approaches: A Framework and Application', *International Journal of Production Economics,* 43: 193–211.

Wagner, M. (2013), '"Green" Human Resource Benefits: Do they Matter as Determinants of Environmental Management System Implementation?', *Journal of Business Ethics,* 114: 443–456.

Walley, N. and Whitehead, B (1994), 'It's Not Easy Being Green', *Harvard Business Review,* May–June: 46–52.

Wang, Q. (2006), 'China's Energy End-Use and Efficiency by International Standards', *Energy of China,* 28(12).

WBCSD (1996), Ceres Initiative launch programme on financial indicators of sustainable development, *Financing Change,* World Business Council for Sustainable Development.

WBCSD (2010), *Vision 2050: The new agenda for business,* WBCSD0167 World Business Council for Sustainable Development.

Welch, E. W., Mori, Y. I., Aoyagi-Usui, M. (2002), Voluntary Adoption of ISO 14001 in Japan: Mechanisms, Stages and Effects, *Business Strategy and the Environment,* 11(1): 43–55.

Welford, R. (1996), *Corporate Environmental Management,* London, Earthscan Publications.

Welford, R. and Gouldson, A. (1993), *Environmental Management and Business Strategy,* London, Pitman Publishing.

Wheeler, D. (1993), *Building Ecological and Public Attitude Concerns into your Business Strategy,* Conference: World Class Corporate Planning, IIR, 23–25 November.

Wibeck, V. (2012), 'Images of Environmental Management: Competing Metaphors in Focus Group Discussions of Swedish Environmental Quality', *Environmental Management,* 49, 776–787.

Winn, M. I. and Angell, A. C. (2000), 'Towards a Process Model of Corporate Greening', *Organizational Studies,* 21(6): 1119–1147.

World Low Carbon and Eco-economy Conference and Technical Exposition (2009), Low-Carbon Economy, English.nc.gov.cn/onlineservice/.../W020110412384019686029.doc.

Yin, H. and Schmeidler, P. J. (2008), 'Why Do Standardized ISO 14001 Environmental Management Systems Lead to Heterogeneous Environmental Outcomes?', *Business Strategy and the Environment,* 18(7): 469–486.

Zadek, S. (2001), *The Civil Corporation: the New Economy of Corporate Citizenship,* Earthscan, London.

Zeng, S. X., Tian, P. and Shi, J. J. (2005) 'Implementing integration of ISO 9001 and ISO 14001 for construction', *Managerial Auditing Journal,* 20(4): 394–407.

INDEX